Essays in Biochemistry

Essays in Biochemistry

Edited for The Biochemical Society by

P. N. Campbell
Department of Biochemistry
University of Leeds
England

F. Dickens
15 Hazelhurst Crescent
Findon Valley
Worthing, Sussex,
England

Volume 8

1972

Published for The Biochemical Society by Academic Press
London and New York

ACADEMIC PRESS INC. (LONDON) LTD
24/28 Oval Road
London NW1

U.S. Edition published by

ACADEMIC PRESS INC.
111 Fifth Avenue
New York
New York 10003

Library of Congress Catalog Card Number: 65-15522
ISBN 0-12-158108-X

Printed in Great Britain by
William Clowes & Sons Limited
London, Colchester and Beccles

List of Contributors

JAMES BADDILEY, *Microbiological Chemistry Research Laboratory, School of Chemistry, University of Newcastle upon Tyne, Newcastle upon Tyne NE1 7RU, England* (p. 35)

P. F. KNOWLES, *Astbury Department of Biophysics, University of Leeds, Leeds LS2 9JT, England* (p. 79)

H. A. KREBS, *Metabolic Research Laboratory, Nuffield Department of Clinical Medicine, Radcliffe Infirmary, University of Oxford, Oxford OX2 6HE, England* (p. 1)

C. Y. LAI, *Roche Institute of Molecular Biology, Nutley, New Jersey 07110, U.S.A.* (p. 149)

G. H. LATHE, *Department of Chemical Pathology, University of Leeds, Leeds LS2 9NL, England* (p. 107)

B. L. HORECKER, *Roche Institute of Molecular Biology, Nutley, New Jersey 07110, U.S.A.* (p. 149)

Preface

On a recent visit to Australia one of us was reminded by a reader that we aimed to produce a series of essays that could be read with pleasure in bed. (Preface to Vol. 5). It seems that one reader at least felt that we had failed in this respect and that some of the essays in recent volumes had been too meaty. While I hastened to point out that there were many reasons why, in that beautiful country, someone might find difficulty in concentrating on essays while reclining in bed, the editors had already given thought to this matter. One of the original intentions of the publication was to avoid comprehensive reviews and encourage a more personal approach.

We hope that our readers will agree that we have been very fortunate in the authors of the present volume and that some at least of the essays may well divert the attention of our Australian readers from other matters. It is no secret that we have been "after" Sir Hans for many years and on various occasions felt really certain that we had "salt on his tail" only to be disappointed because with advancing years he seemed to get even busier. Now we are grateful for a fine essay that we are sure will be much appreciated by our readers of all ages.

The contribution by Dr Peter Knowles on the newer methods of spectroscopy is an attempt to provide a thumb-nail sketch of the potential of the applications of these newer techniques to biochemistry. If this contribution is well received we hope that we may include in future volumes introductions to other techniques and we would like to have suggestions in this respect.

As always we are grateful to our contributors who have once again received our suggestions with good humour. We also appreciate very much the ideas for future essays which we constantly receive and would like to hear from readers who may be on smaller islands than Australia.

June 1972

P. N. CAMPBELL
F. DICKENS

Conventions

The abbreviations, conventions and symbols used in these Essays are those specified by the Editorial Board of *The Biochemical Journal* in *Policy of the Journal and Instructions to Authors* (Revised 1972). The following abbreviations of compounds, etc., are allowed without definition in *The Biochemical Journal.*

ADP, CDP, GDP, IDP, UDP: 5'-pyrophosphates of adenosine, cytidine, guanosine, inosine, uridine and xanthosine
AMP, etc.: adenosine 5'-phosphate, etc.
ATP, etc.: adenosine 5'-triphosphate, etc.
CM-cellulose: carboxymethylcellulose
CoA and acyl-CoA: coenzyme A and its acyl derivatives
DEAE-cellulose: diethylaminoethylcellulose
DNA: deoxyribonucleic acid
Dnp-: 2,4-dinitrophenyl-
Dns-: 5-dimethylaminonaphthalene-1-sulphonyl-
EDTA: ethylenediaminetetra-acetate
FAD: flavin-adenine dinucleotide
FMN: flavin mononucleotide
GSH, GSSG: glutathione, reduced and oxidized
NAD: nicotinamide-adenine dinucleotide
NADP: nicotinamide-adenine dinucleotide phosphate
NMN: nicotinamide mononucleotide
P_i, PP_i: orthophosphate, pyrophosphate
RNA: ribonucleic acid
TEAE-cellulose: triethylaminoethylcellulose
tris: 2-amino-2-hydroxymethylpropane-1,3-diol

The combination NAD^+, NADH is preferred.

The following abbreviations for amino acids and sugars, for use only in presenting sequences and in Tables and Figures, are also allowed without definition.

Amino acids

Ala: alanine
Arg: arginine
Asn*: asparagine
Asp: aspartic acid
Asx: aspartic acid or asparagine (undefined)
Cys: cysteine
CyS: cystine (half)
Gln†: glutamine
Glu: glutamic acid

Glx: glutamic acid or glutamine (undefined)
Gly: glycine
His: histidine
Hyl: hydroxylysine
Hyp: hydroxyproline
Ile: isoleucine
Leu: leucine
Lys: lysine
Met: methionine

Orn: ornithine
Phe: phenylalanine
Pro: proline
Ser: serine
Thr: threonine
Trp: tryptophan
Tyr: tyrosine
Val: valine

* Alternative, $Asp(NH_2)$ † Alternative, $Glu(NH_2)$

Sugars

Ara: arabinose	Glc*: glucose
dRib: 2-deoxyribose	Man: mannose
Fru: fructose	Rib: ribose
Gal: galactose	Xyl: xylose

* Where unambiguous, G may be used.

Abbreviations for nucleic acid used in these essays are:

mRNA: messenger RNA
nRNA: nuclear RNA
rRNA: ribosomal RNA
tRNA: transfer RNA

Any other abbreviations are given on the first page of the text.

References are given in the form used in *The Biochemical Journal,* except that the last as well as the first page of each article is cited and also the title. Titles of journals are abbreviated in accordance with the system employed in the *Chemical Abstracts Service Source Index* (1969) and its Quarterly Supplement (American Chemical Society).

Enzyme Nomenclature

At the first mention of each enzyme in each Essay there is given, whenever possible, the number assigned to it in *Enzyme Nomenclature: Recommendations (1964) of the International Union of Biochemistry on the Nomenclature and Classification of Enzymes, together with their Units and the Symbols of Enzyme Kinetics,* Elsevier Publishing Co., Amsterdam, London and New York, 1965; this document has also appeared as Vol. 13 (2nd edn, 1965) of *Comprehensive Biochemistry,* (Florkin, M. & Stotz, E. H., eds.). Elsevier Publishing Co., Amsterdam, London and New York. Enzyme numbers are given in the form EC 1.2.3.4. The names used by the authors of the Essays are not necessarily those recommended by the International Union of Biochemistry.

Contents

The Pasteur Effect and the Relations Between Respiration and Fermentation*

H. A. KREBS

*Metabolic Research Laboratory, Nuffield
Department of Clinical Medicine, Radcliffe Infirmary,
University of Oxford, Oxford OX2 6HE, England*

* In this essay the terms "glycolysis" and "lactic acid fermentation" (or "fermentation") are used synonymously. The term respiration refers to consumption of oxygen.

I. Early History

The discovery that oxygen inhibits fermentation (or, what amounts to the same thing, that the rate of fermentation rises when oxygen is excluded) was reported by Pasteur in 1861[1]. His observations referred in the first instance to alcoholic fermentation and to lactic acid fermentation. Pasteur himself never commented on the mechanism by which oxygen suppresses fermentation. He was satisfied to look upon the phenomenon as a feature of life which in his times did not seem approachable in terms of chemistry and physics. It was in this frame of mind that he coined[2] his famous saying "Fermentation is life without oxygen".*

Later investigators who observed the accumulation of lactate in muscle under anaerobic conditions and the removal of the lactate on admission of oxygen[3,4,5] assumed that the removal of the products of fermentation was due to their oxidation. The erroneous nature of this assumption was established in 1920 by the work of Meyerhof[6,7] who was the first to make the required quantitative measurements. He measured the rate of oxygen consumption and the rate of lactate removal, working with frog muscle, and found that the amount of oxygen taken up was far too small, by a factor of up to 6, to account for the complete oxidation of lactate. One molecule of oxygen caused the disappearance of two molecules of lactate; for the complete oxidation of this amount of lactate 6 molecules would be needed. At that time it was already established that lactate formation from carbohydrate was a process which can supply energy for muscular contraction. Meyerhof went a step further and assumed that lactate production was *the* energy source in muscle, even aerobically. He thought that the lactate formed by a primary series of reactions was removed by a resynthesis of carbohydrate at the expense of the energy produced by respiration. He visualized a "lactate cycle" involving an anaerobic formation of lactate from carbohydrate and an aerobic resynthesis of carbohydrate:

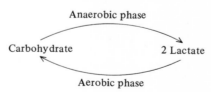

An essential aspect of this new development was the realization that it is not the mere presence of oxygen but the utilization of oxygen which causes the inhibition of fermentation.

---

* This often quoted sentence is a contraction of the following actual text: "*La fermentation est la conséquence de la vie sans gaz oxygène libre . . . L'acte chimique de la fermentation est essentiellement un phenomène corrélatif d'un acte vital, commençant et s'arrêtant avec ce dernier*".

Analogous findings were made subsequently in the 1920's by Meyerhof[5] and by Warburg[8] when they measured the aerobic and anaerobic formation of lactate in various tissues and compared it with the oxygen consumption, using Warburg's new manometric methods. On average the consumption of one molecule of oxygen decreased lactate production by two molecules. The same kind of quantitative relations in this "inhibition of fermentation by respiration" were also established for alcoholic and lactic fermentations of microorganisms. The suppression of fermentation by respiration was found in virtually every type of cell which can ferment sugar and can respire, including plants.[9]

II. Warburg's Experiments with Ethylcarbylamine

An important conceptual development was Warburg's discovery[10] in 1926 that the link between respiration and fermentation could be severed by a specific inhibitor, ethylcarbylamine (ethyl isocyanide, C_2H_5NC). This substance, he found, although not inhibiting the respiration of many cells, nevertheless caused them to glycolyse in the presence of oxygen at almost the same rate as in its absence. In other words, ethylcarbylamine stimulates *aerobic* glycolysis. On the basis of these experiments with ethylcarbylamine, Warburg visualized a link between respiration and fermentation in terms of a specific chemical reaction; he introduced the term "Pasteur Reaction" for this chemical link. This link was taken to be responsible for what subsequently came to be called the "Pasteur effect". He looked upon ethylcarbylamine as a substance which specifically inhibits the Pasteur reaction, while at certain concentrations it does not affect other enzymes. Since ethylcarbylamine, being a derivative of hydrocyanic acid, readily chelates with heavy metals, Warburg suggested that the catalyst promoting the Pasteur reaction may contain a heavy metal. Thus, although the effect of oxygen on glycolysis had been discovered by Pasteur in 1861, it was not named "Pasteur effect" until Warburg introduced the term in 1926. The nature of the hypothetical "Pasteur reaction" and its mechanism of action, however, remained obscure for almost 40 years in spite of numerous efforts to elucidate it (for reviews see Burk[11,12]).

III. Lipmann's "Redox" Hypothesis

In 1933 Lipmann[13,14] discussed what he considered to be a model of the Pasteur effect in glycolysing muscle extracts. Addition of redox indicators which changed the redox state in the direction of oxidation decreased the rate of lactic acid production in muscle extracts and he therefore suggested that the effect of oxygen might be connected with a change of the redox state in the same direction. He used indophenols, the standard potential of which is of the order of +100 mV. Indophenols are reduced anaerobically by tissue extracts but are re-oxidized by molecular oxygen and maintain a redox potential in the solution of about +100 mV. Lipmann's hypothesis was plausible but at that time, it should be noted, NAD had not yet been discovered and the mechanism of the oxido-reductions concerned in fermentation was entirely obscure. It is now established that the redox potential of the NAD system in glycolysing systems is between -200 and -240 mV.[15] At this potential the ratio of [free NAD^+]/[free NADH] is of the order of 1000. At a potential of +100 mV the value of the above ratio would be 10^{14} and the concentration of free NADH would be exceedingly small. This would preclude the reductive formation of lactate from pyruvate. Furthermore the changes of the redox state of the NAD couple in the cytoplasm which may occur under physiological conditions are relatively small; this invalidates Lipmann's hypothesis.

IV. Inorganic Phosphate as a Controlling Agent

In 1941, Lynen[16] and Johnson[17] independently suggested that inorganic phosphate may be a key substance in controlling the rate of glycolysis because it is a reactant in the conversion of glyceraldehyde phosphate to phosphoglyceric acid. At that time it was already known that the concentration of phosphate can be decreased by oxidative phosphorylation and it was therefore logical to postulate that respiration and fermentation compete for inorganic phosphate. In the presence of oxygen, respiration would win in this competition and therefore decrease the rate of glycolysis. It was a weakness of this concept that it did not explain the decreased utilization of carbohydrate. The step which requires inorganic phosphate is many stages removed from the initial reaction of fermentation. So while the hypothesis would explain the non-fermentation it could not account for the cessation of the initiation of fermentation. Another weakness was the fact that the concentration of inorganic phosphate in many materials does not change sufficiently on the transition from anaerobic to aerobic conditions to explain the differences in the rate of glycolysis, even though there are some tissues[18] where the changes may be significant.

V. Uncoupling of Oxidative Phosphorylation

A decisive development which eventually made a major contribution to the elucidation of the Pasteur effect was the clarification of the nature of action of nitrophenols on cell metabolism. Nitrophenols had been known since 1917 to stimulate biological oxidations, but this information remained buried in the literature on industrial toxicology until 1932. The discovery was made by André Mayer[19] and his colleagues in Paris when they investigated a severe illness of factory workers who handled picric acid (trinitrophenol), manufactured on a large scale because it was needed as an explosive during World War I. One of the main causes of death by picric acid poisoning was hyperpyrexia caused by increased rates of cellular combustion. It was not until 1932 that Magne, Mayer and Plantefol[19] published a detailed account of the toxic properties of nitrophenols. This was followed by the discovery by Dodds and Greville[20] that nitrophenols, although increasing respiration, caused many cells and tissues to glycolyse aerobically. In this respect the action of nitrophenols was similar to that of ethylcarbylamine. In 1936, Clowes and Krahl[21] found that nitrophenols also inhibit cell division and in 1940 Clifton[22] established the molecular basis of this inhibition: nitrophenols in general inhibit biosynthetic processes. As early as 1942, on the basis of these observations, Lipmann[14] expressed the view that nitrophenols may be useful tools in the study of the Pasteur effect. The culmination of this development was the discovery by Loomis and Lipmann[23] in 1948, that nitrophenols uncouple phosphorylation from respiration; they prevent the formation of ATP in respiring systems. This at once explained the inhibition of biosynthetic reactions as these depend on a supply of ATP. At that stage of knowledge, however, the nature of the Pasteur reaction was still obscure and in fact the full relevance of the discovery of the uncoupling effect to the explanation of the Pasteur effect, was not, and at that time could not be, appreciated.

VI. The Identification of Phosphofructokinase as the Site of the Pasteur Effect

The modern development which eventually provided a solution of the nature of the Pasteur effect and the Pasteur reaction was initiated in 1943 by Engelhardt and Sakov,[24] although this work was no more than a beginning. Engelhardt and Sakov reasoned that the Pasteur effect must operate at the stage of glycolysis when fructose 6-phosphate is phosphorylated by ATP, their line of reasoning being as follows: the effect has to be explained on the basis of an inhibition of a specific stage of glycolysis. Since Engelhardt and Sakov believed that glucose 6-phosphate was a stage in what was called the oxidative shunt pathway of glucose degradation, they believed that the formation of glucose 6-phosphate could not be inhibited under aerobic conditions. Furthermore, since fructose 1,6-diphosphate was readily fermented in the presence of agents which inhibited *glucose* fermentation, none of the reactions involved in the fermentation of fructose diphosphate could be the point of inhibition by oxygen. So by eliminating other stages they concluded that phosphofructokinase must be the point where the Pasteur effect operates. The limited general knowledge on the regulation of enzyme activity prevented Engelhardt and Sakov from commenting in detail on the mechanism of this inhibition.

The position in 1951 was critically reviewed by Dickens.[25] His article gives a very fair picture of the numerous relevant observations and the attempts to explain the Pasteur reaction as they appeared to the critical reviewer at that time.

The final solution began to emerge in the later 1950's. Retrospectively, it can be said that it had two roots. One was the detailed study of the properties of phosphofructokinase, in particular of the factors which activate and inactivate the enzyme. The other was the formulation of the general concept of "allosteric" properties of enzymes, i.e. of the realization that enzymes are not merely catalysts but often possess built-in control mechanisms which regulate their activity.

The postulate of Engelhardt and Sakov was further supported by other studies, based essentially on the measurement of the changes in the concentrations of intermediates of glycolysis, all of which pointed to phosphofructokinase as the site of the Pasteur effect. Such observations were reported around 1960 by Lynen[26] on yeast cells, by Lonberg-Holm[27] on ascites tumour cells, by Park et al.[28] on heart muscle and by Newsholme and Randle[29] on diaphragm.

TABLE 1

Substances affecting phosphofructokinase activity (Passonneau and Lowry[34]).

Inhibitors	Enhancers	De-inhibitors of ATP, citrate or Mg^{2+}
ATP	NH_4^+	Fructose diphosphate
Citrate	K^+	cyclic 3', 5'-AMP
Mg^{2+}	P_i	5'-AMP
	5'-AMP	ADP
	cyclic 3',5'-AMP	Fructose 6-phosphate
	ADP	P_i
	Fructose diphosphate	

Earlier observations (1944) by Iwakawa[30] on skeletal muscle can also be interpreted as support. Bücher[31] came to the conclusion that in insect flight muscle the activity of phosphofructokinase rises during activity and that the concentrations of ATP, Mg^{2+} and fructose 6-phosphate regulate the activity of phosphofructokinase. In 1962, Mansour and Mansour[32] showed that phosphofructokinase of the liver fluke can also be activated by ATP, Mg^{2+} and cyclic AMP, and Passonneau and Lowry[33] made similar observations on muscle phosphofructokinase. Their work added fructose diphosphate as a powerful activator to the list of substances which control the activity of phosphofructokinase. Other activators are K^+ and NH_4^+. Two kinds of activators of phosphofructokinase can be distinguished; enhancers which increase the activity of the enzyme in the absence of inhibitors and de-inhibitors which counteract the effects of inhibitors. Passonneau and Lowry[34] classified the effectors of phosphofructokinase as shown in Table 1.

Further work by Wu,[35] Vinuela, Salas and Sols,[36] Garland, Randle and Newsholme,[37] Parmeggiani and Bowman,[38] Mansour, Wakid and Sprouse[39] (who first crystallized the enzyme), Ahlfors and Mansour,[40] Paetkau and Lardy,[41] Hulme and Tipton,[42] also by Ho and Anderson,[43] Ibsen and Schiller,[44] Mansour,[45] and Krzanowski and Matschinsky,[46] established the fact that phosphofructokinase from many different sources, from microorganisms to higher animals, exhibits many similar, though not absolutely identical, regulatory characteristics. Phosphocreatine was added to the list of inhibitors by Krzanowski and Matschinsky.[46] Mansour[47] has presented evidence indicating that the reversible formation of an inactive form of phosphofructokinase may also play a regulatory role.

VII. Inhibition of Hexokinse

The allosteric properties of phosphofructokinase can explain an inhibition of glycolysis at the stage of fructose 6-phosphate but they do not account for the non-utilization of *glucose*, when the Pasteur effect operates. Thus an additional factor must play a role. Many observations fit in with the assumption that this is the inhibition of hexokinase by glucose 6-phosphate, first described for brain hexokinase by Weil-Malherbe and Bone[48] and Crane and Sols.[49] The inhibition is non-competitive and can be pronounced at 0·5 mM. Hexokinase can also be inhibited by ADP and 5'-AMP,[50,51] but experiments by Randle, Denton and England[52] show that, at least in the perfused rat heart, the concentration of glucose 6-phosphate is the main factor responsible for the regulation of hexokinase activity.

When phosphofructokinase is inhibited, the substrate of this enzyme, fructose 6-phosphate, accumulates. Owing to the high activity of hexosephosphate isomerase the accumulation of fructose 6-phosphate is coupled with an accumulation of glucose 6-phosphate. Thus inhibition of phosphofructokinase automatically generates the inhibitor of hexokinase. Between them, the allosteric properties of phosphofructokinase and the inhibition of hexokinase by glucose 6-phosphate satisfactorily explain the main features of the Pasteur effect. Numerous experiments have been recorded which show that the postulated changes in the concentrations of the effectors of phosphofructokinase and of glucose 6-phosphate correlate to the rate of glycolysis or of fermentation.[52,53,54,55]

Hexokinase, it may be argued, also initiates the pentose phosphate pathway; why, then, does a rise in the concentration of glucose 6-phosphate not accelerate also the glucose 6-phosphate dehydrogenase reaction and the pentosephosphate cycle? There is no satisfactory answer to this question. It has been postulated that a specific inhibitor, not yet identified, must operate as a controlling factor of this enzyme and of the pentose phosphate cycle.[56,57] The cycle is furthermore regulated by enzymic adaptation. The capacity of glucose 6-phosphate dehydrogenase can fall to one tenth of the maximum value when the requirement for $NADPH_2$ is minimal, as is the case in the absence of fatty acid synthesis.[58]

VIII. The Reasons for the Multiplicity of Effectors of Phosphofructokinase

Physiological considerations can be put forward to explain the unusual multiplicity of effectors of phosphofructokinase. When glucose is a main source of energy, undergoing either oxidation or fermentation (depending on the availability of oxygen), the phosphorylation states of the adenine nucleotides, i.e. the concentrations of ATP, ADP, and P_i are ideal signals for the control of glucose degradation. In this situation, the maintenance of a high concentration of ATP by resynthesis from ADP and P_i is the physiological reason for glucose degradation. There are other situations where it is essential that glucose degradation is inhibited even when the concentration of ATP is relatively low. This applies for example to conditions when fatty acids are the main fuel of respiration, as is the case in starvation, or on a diet low in carbohydrate and rich in fat. In this situation fatty acids supply the required energy, and glucose must not be degraded even though the concentration of ATP may be low and that of ADP high. Here the inhibition of phosphofructokinase by citrate is a major regulatory factor.[52]

There is another physiological situation where glycolysis must occur even though it does not primarily serve to supply energy, namely when carbohydrate is converted to fat in the liver or in the adipose tissue. Up to the stage of the mitochondrial formation of acetyl CoA the pathways of fatty acid synthesis from glucose and of the combustion of glucose are identical, and a high rate of glycolysis must therefore be maintained during lipogenesis. This is achieved by the diversion from the mitochondria to the cytosol of the citrate formed from acetyl CoA and oxaloacetate. In this compartment citrate is split by citrate lyase to acetyl CoA and oxaloacetate with an expenditure of ATP; whilst the acetyl CoA, again with the expenditure of ATP, is synthesized to fatty acids. Thus fatty acid synthesis entails an extra expenditure of ATP and thereby maintains conditions favourable for high activity of phosphofructokinase. Inhibition of this enzyme by citrate is prevented by the high activity of the inducible citrate lyase.

The significance of the activation of phosphofructokinase by cyclic AMP, and by NH_4^+, cannot be formulated with confidence. The activation of phosphofructokinase by its product, fructose 1,6-diphosphate, belongs to the category of positive feedback mechanisms and co-operative activation, not uncommon phenomena among regulatory enzymes.

IX. Aerobic and Anaerobic Removal of Fructose

In many animal tissues fructose can replace glucose as a source of energy,[59,60,61,62] and in the majority of these tissues the utilization of fructose is initiated by the same hexokinase as is responsible for glucose utilization.[63] When fructose is the substrate of this enzyme the primary product is fructose 6-phosphate, and the mechanisms of the regulation of fructose and glucose degradation are therefore identical.

A few tissues, however—especially liver, kidney cortex and the mucosa of the gut[64,65,66,67,68,69]—contain a second enzyme which can initiate fructose utilization. The pathway and the enzymes involved are as follows:

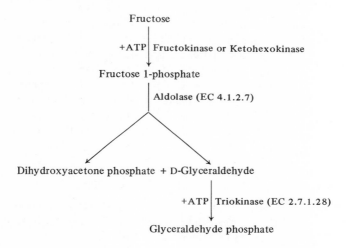

This pathway is not subject to the regulatory controls of the hexokinase pathway, as it does not involve hexokinase or phosphofructokinase. In consequence there is no Pasteur effect as far as fructose utilization in liver, kidney cortex and intestinal mucosa is concerned. What, then, is the significance of the above reactions? The following considerations attempt to answer this question.

The synthesis of hepatic glycogen from glucose or fructose necessitates the formation, and a sufficient rise in the steady state concentration, of glucose 6-phosphate. This rise causes an inhibition of hexokinase so that this enzyme cannot play a major part in the formation of hepatic glycogen. The role of hexokinase as far as glucose utilization is concerned, is taken by glucokinase, an enzyme catalysing the same reaction as hexokinase, but only at relatively high glucose concentrations as its K_m is about a thousand times higher than that of hexokinase.[63] The existence of this enzyme (which is inducible and highly active only when there is an excess of carbohydrate to be deposited) makes the formation of glucose 6-phosphate independent of the regulatory mechanisms which control glucose degradation at low glucose concentrations. But glucokinase, unlike hexokinase, does not react with fructose and the synthesis of glycogen from fructose therefore requires another enzymic mechanism which leads to glucose 6-phosphate. This pathway is provided by the reactions of the scheme on page 9 followed by the formation of fructose diphosphate, fructose 6-phosphate and glucose 6-phosphate. This concept explains a series of long established facts, the significance of which has been somewhat puzzling; the existence of *two* pathways of fructose utilization, involving enzymes of very high capacity such as fructokinase and triokinase (which converts glyceraldehyde and dihydroxyacetone into their phosphate esters), and the very high rates at which the triosephosphates and the trioses can be converted to glucose.

To sum up, the special metabolic reactions of fructose just discussed are, physiologically speaking, the equivalent of glucokinase, permitting glycogen synthesis when regulatory mechanisms block the normal conversion of fructose to its 6-ester which initiates the degradation of fructose. This interpretation is supported by an analysis of the kinetic properties of triokinase[70] and by the inducibility by fructose of fructose 1-phosphate aldolase and of triokinase.[71]

X. Energetics of the Pasteur Effect

Meyerhof[5] and Warburg[8] established empirically that the consumption of 1 molecule of O_2 usually prevents the formation of 2 molecules of lactate, or (in yeast cells) of 2 molecules of ethanol and carbon dioxide. As a quantitative measure of the effect of oxygen on fermentation, Warburg introduced what he termed the Meyerhof quotient which he defined as follows:

$$\text{Meyerhof quotient} = \frac{\text{anaerobic fermentation} - \text{aerobic fermentation}}{O_2 \text{ uptake}}$$

When the rate of fermentation is expressed in terms of amounts of lactic acid (or ethanol or CO_2) per unit time and in the same units as the O_2 consumption, a value of 2 is usually found, though variability is considerable. A value of 2 cannot be explained on the basis of simple energetics. The utilization of one molecule of O_2 yields up to 6 molecules of ATP whilst the formation of one molecule of lactate or ethanol yields only 1 molecule of ATP. So on the basis of energetics a value of 6 would be expected for the Meyerhof quotient. Values up to 5 have in fact been found in ascites tumour cells,[72] but this is an exceptional case. It follows that fermentation as a rule does not fully replace respiration as far as the rate of supply of utilizable energy is concerned.

The allosteric inhibition of phosphofructokinase and of hexokinase, in view of the exceedingly low enzyme concentrations, should require negligible amounts of ATP, and energetically the inhibition of phosphofructokinase is therefore very much cheaper than a resynthesis of glucose from lactate as originally visualized by Meyerhof. This resynthesis is known to require 6 molecules of ATP per molecule of glucose formed.[73] Thus a Meyerhof quotient of 2 would be sufficient for the resynthesis of one molecule of glucose from lactate. Such a resynthesis occurs in liver and kidney cortex, but the extra oxygen consumption measured under conditions of gluconeogenesis from lactate is substantially greater than expected.[74]

The Pasteur effect and the value of the Meyerhof quotient, then, are not matters of stoicheiometry or simple energetics. The energetic aspects may be looked upon as follows. Two types of energy-supplying processes—oxidation by O_2 (respiration) and the O_2-independent fermentations—are available to the majority of living cells. In terms of energy yields respiration is always more economic than fermentation. Thus one molecule of glucose can form 38 molecules of ATP when oxidized, but only 2 when fermented. Moreover fermentation causes an accumulation of harmful end-products which create an environment unfavourable for life. Lactic acid formation shifts pH to an inhibitory range; ethanol at higher concentrations has narcotic effects which paralyse cell activities generally. Thus the replacement of fermentation by respiration has a great biological advantage and it is therefore not surprising that a mechanism has evolved which suppresses fermentation in the presence of O_2 when fermentation fulfils no useful function. So the Pasteur effect, from the physiological point of view, means that respiration replaces fermentation as a more economic, and potentially less harmful, process. The fact that the Meyerhof quotient as a rule is much below the theoretically expected value of 6 may be simply due to the circumstance that the capacity for anaerobic fermentation is usually too low, and the respiratory capacity too high to make the suppression of fermentation by respiration energetically equivalent.

Though the Meyerhof quotient can no longer be regarded as an indicator of the "efficiency" of the Pasteur effect, it is still a useful characteristic. In the light of the newer concepts it has acquired a new meaning: it indicates whether or not an increase in ATP supply and ATP expenditure occurs on transition from anaerobic to aerobic conditions. It follows from what has already been stated that a Meyerhof quotient of 6 (rarely realized) would mean that the aerobic and anaerobic energy supply would be the same, on the assumption—only an approximation—that 6 molecules of ATP are formed per molecule of O_2 used. A lower value of the Meyerhof quotient indicates an increase in ATP supplying and requiring processes on transition from anaerobic to aerobic conditions. On the assumption that the aerobic fermentation is zero, the ATP supply is increased by a factor approximately equal to

$$\frac{6}{\text{Value of the Meyerhof quotient}}$$

Thus the increase in ATP supply is 6-fold if the Meyerhof quotient is one, 4-fold if the Meyerhof quotient is 1·5 and 3-fold if the Meyerhof quotient is 2. The formula does not hold when the aerobic fermentation is significant.

XI. The Physiological Role of Aerobic Glycolysis in Various Animal Tissues

The ability to obtain energy anaerobically is obviously an advantage in the absence of O_2; but if respiration is a more effective way of obtaining energy, why, then, are there animal tissues which normally obtain some energy from glycolysis in the presence of oxygen? Such tissues are striated muscle, intestinal mucosa, renal medulla, erythrocytes, foetal tissues during parturition, malignant tumours and retina.

First, two general comments: (1) The wastefulness of glycolysis is minimized in the animal body because the lactate formed can eventually be utilized, either directly as a fuel, or after resynthesis to glucose in liver or kidney; (2) The physiological advantages of aerobic glycolysis vary from tissue to tissue, as the following considerations indicate.

A. STRIATED MUSCLE

In severe exercise the oxygen supply by the circulation limits the rate of respiration and lactic acid formation from glycogen stores can supply extra ATP. Glycolysis is essential for maximum physical work.[75,76]

B. INTESTINAL MUCOSA

The aerobic formation of lactate by the intestinal mucosa, discovered by Dickens and Weil-Malherbe[77] in 1941, may play a role in the absorption of glucose and fructose from the gut because the lactate formed within the intestinal mucosa is unidirectionally discharged into the blood circulation, as opposed to the lumen of the intestine (for review see[78,79]). This was first observed on isolated everted sacs of the gut and has been confirmed by the determination of lactate in portal blood. In the case of fructose as much as 50% of the absorbed sugar appeared as lactate.[80] In the case of glucose the percentage is lower. The key glycolytic enzymes, especially fructokinase, but also glucokinase and hexokinase, of the intestinal mucosa are adaptive. Their capacities decrease on fasting and increase on feeding glucose or fructose.[81,82,83]

C. RENAL MEDULLA

The relatively high capacity for aerobic glycolysis of the inner renal medulla even in serum, first reported by Dickens and Weil-Malherbe,[84] is taken to be an important source of energy in the counter-current mechanism of urine concentration.[85,86] In some species (guinea-pig, cat) the glycolytic capacity per unit tissue weight is one of the highest of mammalian organs, exceeded only by the muscles, blood cells and retina. A high rate of aerobic glycolysis, first established for slices, has been confirmed for the intact tissue by Ruiz-Guinazu, Pehling, Rumrich and Ullrich[87] who used the micropuncture technique and compared the glucose and lactate concentrations in the renal artery and the vasa recta. They found a fall in the glucose concentration and a rise in the lactate concentration. It may be relevant that *in vivo* oxygenation of the medulla, as indicated by its white colour and the low haematocrit value of the blood in the vasa recta,[88,89] is limited. However, the rate of glycolysis is also high in well-oxygenated slices. The physiological advantage to the renal medulla of the energy supply by glycolysis is not apparent.

D. ERYTHROCYTES

Glycolysis is known to provide the energy for maintaining the physiological environment of the mammalian red cell.[90] But why mature mammalian erythrocytes have lost their respiratory capacity while the nucleated red cells of other vertebrates obtain their energy from respiration is another open question. Any attempt to explain the fact that mammalian red cells glycolyse aerobically must take into account the fact that the red cells of other vertebrates respire instead. Warburg's assumption that the non-nucleated mammalian red cells are "dying" cells has not proved correct because the life span of avian erythrocytes is, if anything, even shorter than that of mammalian erythrocytes.[91]

It is likely that the reactions of glycolysis are of special importance for the supply of 2,3-diphosphoglycerate which occurs in high concentrations (up to 5 mM) in mammalian red cells, mainly in combination with haemoglobin. It has recently come to light that it plays an important part in the regulation of the O_2 binding capacity of haemoglobin at lower O_2 pressures. It is a cofactor of oxygen unloading because by combining with haemoglobin it decreases the latter's affinity for O_2.[92] 2,3-Diphosphoglycerate is formed and further metabolized by a bypass of glycolysis, catalysed by diphosphoglycerate mutase and diphosphoglycerate phosphatase:[93]

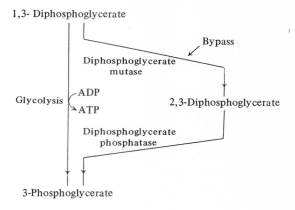

Thus glycolysis is essential for the formation of an important constituent of the red cell. Moreover the concentration of 2,3-diphosphoglycerate is not constant; it changes as an adaptive measure, according to the physiological needs, increasing for example in various forms of hypoxia when a low affinity of O_2 for haemoglobin at low O_2-pressures is vital.[94,95]

In the present context it is very relevant that the red cells of non-mammalian vertebrates do not contain much 2,3-diphosphoglycerate. Its place as co-factor of oxygen release is in these cells taken by inositol hexaphosphate,[96,97] a substance discovered in nucleated erythrocytes by Rapoport in 1940.[98] Inositol hexaphosphate is in all probability formed from glucose and ATP.[99] Thus the reactions of glycolysis are not involved and whilst the synthesis of diphospho-glycerate costs one extra ATP that of inositol hexaphosphate probably costs 6 extra ATP. The actual ATP supply needed for maintaining the physiological concentrations of the polyphosphates would depend on their turnover on which there is as yet no detailed information. Meanwhile it is a useful working hypothesis to assume that the absence from non-mammalian vertebrate red cells of aerobic glycolysis and its replacement by a more efficient oxidative ATP generating mechanism may be connected with the differences in the co-factors of oxygen release in the different kinds of red cells.

A connection between red cell glycolysis and 2,3-diphosphoglycerate for-mation is also suggested by the fact that this ester is a powerful inhibitor of red cell hexokinase,[100,101] the first feedback inhibitor ever to be discovered in the field of metabolic pathways (Dische, 1941).[100] This now makes good physiological sense: when diphosphoglycerate accumulates by release from haemoglobin the rate of glycolysis decreases until diphosphoglycerate has been converted to lactate, with the formation of ATP. When diphosphoglycerate is taken up by haemoglobin in hypoxia, glycolysis is accelerated until the conversion of free diphosphoglycerate is restored to its normal level. Without the special function of diphosphoglycerate as a co-factor of O_2 release the significance of the feedback inhibition of hexokinase cannot be properly understood because diphosphoglycerate is not an end-product of glycolysis.

E. FOETAL TISSUES DURING PARTURITION

Large glycogen reserves accumulate in foetal tissues towards the end of term. In the liver the glycogen reserves can rise to 10% of the wet weight.[102] This is matched by a high glycolytic capacity of foetal tissues and the organs of the new-born. In the kidney cortex, for example, it is four times higher in the new-born rat than in the adult rat.[103] Evidence of glycolysis *in vivo* is the high lactate concentration in the tissues of the new-born. Ballard[104] found 5·1 mM lactate in the liver of rats two minutes after birth, against the normal value of adult rats of about 1mM. The glycogen reserves and the high glycolytic capacity have an important function in asphyxia when the infant is separated from the parental sources of energy at birth.[105] The fact that human babies can survive anaerobiosis for 30 min[106,107] is no doubt connected with the high glycolytic capacity of the new-born.

F. MALIGNANT TUMOURS

Ever since Warburg discovered the high glycolytic capacity of malignant tumours it has been evident that glycolysis is an important source of energy in this type of tissue, sufficient even for prolonged survival in the absence of oxygen. Thus transplantable tumours may grow again after having been kept for three days in the absence of oxygen.[108,109] This means that neoplastic tissue shares with other tissues the capacity to obtain energy from two sources. The proportion of energy derived from glycolysis is very high in neoplastic tissue (though it is also high in erythrocytes, renal medulla and retina). Warburg has emphasized that cancer is the only tissue that can grow at the expense of energy supplied by glycolysis. There are of course many other biochemical characteristics which distinguish neoplastic tissue from other tissues.[109a] Between them they are responsible for the uncontrolled growth, but from the quantitative point of view no single characteristic is as striking as the aerobic glycolysis.

G. RETINA

Of special interest is the aerobic glycolysis of the retina of warm-blooded animals, discovered in 1924 by Warburg, Posener and Negelein,[110] because it is exceptionally high, particularly so in the avian retina. The rate of glycolysis of avian retina is in fact the highest found in any animal tissue.[111] The rate of respiration is low in avian retina and the difference between aerobic and anaerobic glycolysis, if any, is very small (Table 2). Warburg suspected the aerobic lactate production to be an artefact, caused by tissue "damage". This

TABLE 2

Respiration and glycolysis of the retina. Values are taken from the literature. The temperature was 37° to 40°C except where stated otherwise. The Q units used are those introduced by Warburg (μl/mg dry wt/h), 1 μl of lactate being taken as equivalent to 22·4 μl gas.

Species	Q_{O_2} Respiration	$Q_L^{O_2}$ Aerobic lactate formation	$Q_L^{N_2}$ Anaerobic lactate formation	Reference
Rat	−31	+45	+88	110
Rabbit	−27	+33	+37	113
Chicken	very small	+109	+105	111
Pigeon	−8	+183	+187	111
Frog (25°C)				
(*Rana esculenta*)	−6	0	+18	112
Fish (20°C)				
(*Leuciscus rutilis*)	−5·5	0·6	+15	116

view was based on the fact that many sensitive tissues glycolyse aerobically when they are exposed to unphysiological conditions. Thus the haemopoietic bone marrow cells which hardly glycolyse when freshly suspended in blood serum, on incubation gradually develop a high aerobic glycolysis while respiration falls (see Table 3). The frog retina does not glycolyse aerobically, but does so when exposed to unphysiological high temperatures.[112] Graymore[113] however thought that the constancy and reproducibility of the aerobic glycolysis of the mammalian retina argued against the assumption of an artefact. Until recently this assumption could not be put to a direct test, because it was not possible to measure the metabolism of the retina *in situ*.

TABLE 3

Changes in respiration and glycolysis of a suspension of haemopoietic cells from rat bone marrow (Fujita[114]). The values are μmol/mg dry wt/h. The cells were suspended in rat serum containing 27 mM bicarbonate and 11 mM glucose in equilibrium with 5% CO_2 in O_2. The figures illustrate the gradual loss of respiration, the relative constancy of anaerobic glycolysis and the gradual rise of aerobic glycolysis on incubation *in vitro*.

Time	O_2 uptake	Lactate production aerobic	anaerobic
First 30 min	0·58	0·17	1·15
Second 30 min	0·49	0·30	1·13
Third 30 min	0·39	0·52	1·01
Fourth 30 min	0·17	0·55	0·93
Fifth 30 min	0·09	0·66	0·92
Sixth 30 min	0	0·78	0·84

Two recent developments have made it possible to approach the problem afresh. Firstly, the enzymic spectrophotometric methods of analysis are now sufficiently sensitive to measure accurately lactate and glucose in a few microlitres of blood, a quantity which can be collected from the veins draining the retina of some species. Secondly, electron microscopy can decide whether or not the retina of birds (which shows hardly any respiration *in vitro*) possesses mitochondria, the respiratory organelles.

Not every species has a vascular system suitable for collecting venous blood samples *in vivo*.[117] The rabbit is a suitable species as the vortex veins draining the eye are readily accessible and the retinal vessels are negligible. Dr R. A. Hawkins and the author[118] have recently collected 0·02 to 0·2 ml venous blood from the superior vortex vein and compared its glucose and lactate content with simultaneous samples of arterial blood from the femoral arteries. The mean arterio-venous difference of glucose was found to be $-0·39$ μmol/ml and that of lactate $+0·37$ μmol/ml. These differences were highly significant ($P < 0·0005$). The rate of blood flow through the chorio-capillaries of the rabbit eye is about 100 ml per eye per hour.[119,120] Thus the lactate production was about 37 μmol per eye per hour, on the assumption that the rate of blood flow in all four vortex veins was the same. The isolated rabbit retina forms aerobically about 30 μmol lactate per eye per hour.[121] A value lower than that given by the whole eye is to be expected because the vortex veins also drain the iris, ciliary body, sclera, choroid, lens, vitreous body and cornea. Between them these form about 10 μmol lactate per eye per hour, a figure based on the measurement of metabolism of the isolated tissues *in vitro*. The lens makes the largest single contribution (4 μmol/lens/h).[122] So the lactate production of the rabbit eye *in vivo* is of the same order (37 μmol) as the calculated sum of lactate formation by the individual tissues measured *in vitro* (40 μmol).

The glucose consumption by the eye would be expected to be a little more than half that of the lactate production, because some glucose might be used as a substrate for oxidation. In fact the observed mean figure of 39 μmol glucose per eye per hour is higher than the expected value, but the difference is entirely within the limits of error. An error of only 1% in the means of the arterial and venous glucose values would account for the difference between the observed and expected value. As the absolute concentration of lactate in the blood was much lower—about 2 mM—than that of glucose (about 8 mM) the error in the determination of lactate was much smaller than that in the determination of glucose and a 1% error would not make a major difference to the lactate measurements.

Since the isolated avian retina, especially that of the pigeon, represents an extreme example of very high glycolysis and very low respiration (Table 2), this tissue is particularly suitable for the examination of the distribution of mitochondria. The literature contains a few comments[123] on the presence of mitochondria in pigeon retina, but there is no information on the quantitative aspects. Dr Trevor Hughes (Department of Neuropathology at the Radcliffe Infirmary, Oxford) has investigated this point by electron microscopy and histochemical methods. As already reported by Cohen,[123] clusters of mito-chondria occur in a small area at the junction of the inner and outer segments of the cones and rods. This area occupies rather less than 4% of the whole retina. It is very close to the zone where light is detected and where its energy is amplified and transformed into a neural impulse, a process which involves ATP.[124] It is the zone close to the pigment cells and the choroid where the oxygen supply from the blood vessels is nearest. All other areas of the pigeon retina show exceedingly small numbers of mitochondria, if any. There are therefore large regions which possess no means of forming ATP by oxidative phosphorylation.

Related to the scarcity of mitochondria is no doubt the complete absence of blood vessels from the avian retina.[117] The area of the most acute vision in the mammalian retina—the macula—it should be recalled, is also free from blood vessels.

There is powerful supporting evidence of another kind in favour of the exist-ence *in vivo* of aerobic lactate production, the very high concentration of lactate in the vitreous body of mammals and birds (Table 4). This was discovered in 1951 by de Vincentiis[125] (Table 4). At that time Warburg's view that retinal glycolysis *in vitro* is probably an artefact dominated thought on the subject and de Vincentiis used guarded terms when he expressed the view that his findings might be taken to support the view of glycolysis also *in vivo*. He remarked that without additional information the assertion that the retina glycolyses *in vivo* "would be rash". And he thought that at best his data are indirect evidence

TABLE 4

Lactate concentrations in vitreous body. Data from the literature: values in ref. 125 refer to plasma; those in ref. 126 to whole blood.

Species	Lactate concentration (mM) Vitreous body	Plasma or blood	Reference
Man	7·7	2·8	125
Rabbit	7·2	2·6	125
Rabbit	8·3-16·7	6·7-14·7	126
Fowl	5·2	1·6	125
Pigeon	16·0	3·5	125

which needed supplementation by the determination of arterio-venous differences. He pointed out that a high concentration of a metabolite in the ocular fluids may be the result of a gradual accumulation caused by a low rate of outflow. The published reports on the high lactate content of vitreous body make no reference to the possibility of post-mortal accumulation of lactate. We have therefore carried out experiments[122] in which the vitreous body was collected *in vivo* in the anaesthetized animal or within seconds of death. The results obtained leave no doubt that a very high lactate concentration occurs *in vivo* (Table 5). Since the glucose concentration in the vitreous body was only slightly lower than that of the blood it cannot have been a postmortal source of lactate.

TABLE 5

Lactate and glucose in the vitreous body and blood. In order to avoid postmortal accumulation of lactate some animals (where indicated) were anaesthetized with nembutal and the vitreous body was collected *in vivo*. The animals were well fed unless otherwise stated. The values for the vitreous body are μmol/g, those for blood μmol/ml. (Krebs, H. A. and Hems, R.; hitherto unpublished observations).

Species	Conditions of collection	Vitreous body Lactate	Glucose	Blood Lactate	Glucose
Pigeon	Nembutal anaesthesia	18-25	9-17	4·0-6·7	14
Rabbit	Nembutal anaesthesia	15-18	2·5-3·0	2·7	3·4
Rat	Killed by stunning. Pooled samples	13	6		
Mouse	Killed by stunning. Pooled samples	18	6		

It is attractive to ascribe a physiological significance to the fact that the retina obtains most of its energy from glycolysis. Vascularization of the retina must somewhat interfere with visual perception because of light absorption by the red blood cells. Mitochondria would decrease the transparency of the retina by light scattering, as well as by light absorption by the cytochromes and flavoproteins.

If the avian retina is an extreme with respect to high glycolysis and low respiration, this correlates to the extreme visual acuity of birds, vital for their survival. During flight high above ground their horizon is very much wider than that of mammals. Vultures, owing to their outstanding long range vision can discover a carcass from enormous distances as they patrol the skies.[127] The buzzard's eye is reported to have an analytical capacity four or five times greater than that of the human eye, one reason being the density of the receptor cells in some parts of the retina. The buzzard sights, watches and follows a mouse from a height from which the human eye could not possibly see it.[128] Evidently, high visual acuity is of decisive survival value to birds, olfactory receptors being rudimentary, or even absent in this class of animal.

The dependence of the avian retina on lactic acid fermentation implies that this tissue requires quantities of glucose which are enormous by comparison with other species. Pigeon retina may consume between 100 and 160 mg glucose per gram wet weight per hour.[111] Brain cortex, by contrast, requires about 4 mg glucose/g/hour and most other tissues need less. In order to meet its glucose requirements the retina of one pigeon eye has to clear the glucose content of about 3 ml blood/h, and pigeon blood (avian blood generally), it should be recalled, has a much higher glucose concentration (10-15 mM) than mammalian blood (5 mM). Were the glucose content of avian blood the same as that of mammalian blood, then 2 to 3 times greater volumes of blood would be needed to feed the retina. Perhaps the high glucose content of avian blood has evolved because it is essential for adequate nutrient supply to the retina. The distance of diffusion from the nearest capillaries in the choroid through the retina to the nerve fibre layer is about 0·3 mm. This distance between the capillaries and the site of nutrient consumption is exceptionally great.

These considerations confirm the conclusion that the high glycolysis of the retina is not an artefact. It has evolved because it confers decisive biological advantages; it contributes to the efficiency of vision. In principle, what has been said about the avian retina must also apply to the mammalian retina, though here the percentage of ATP formed by glycolysis may be only about 20% of the total ATP required.

H. GENERAL COMMENTS

The occurrence *in vivo* of aerobic lactate formation has now been demonstrated for those tissues—renal medulla, intestinal mucosa, foetal tissue and the retina—for which doubts had earlier been expressed about the validity of the *in vitro* measurements. This finally settles the question, often raised by Warburg, whether or not aerobic glycolysis in these tissues is an artefact. He discussed this question again and again[129,112,130] and in 1957[131,132] he still expressed the view that the glycolysis of mammalian erythrocytes and of the retina are only *apparent* exceptions to the rule that high rates of aerobic glycolysis are restricted to malignant tumours. Mammalian red cells, he argued, are not a true exception because they glycolyse only after the loss of the nucleus, when they can no longer multiply and must therefore be looked upon as incomplete and dying cells. He also doubted whether the retina was a true exception and pointed out that the decisive experiment—the estimation of lactate in the ophthalmic veins—still remained to be done. Warburg's view was based on experiments purporting to show that there is no major aerobic glycolysis in the renal medulla if the tissue is suspended in serum reinforced with 11 mM L-lactate (though not with glucose).

The analysis of the examples where lactate is formed aerobically *in vivo* shows, however, that glycolysis serves as an essential source of energy in a variety of different functions, in red cells to maintain structural integrity and, in addition, to supply a key constituent; in renal medulla to supply energy for the concentration of urine; in muscle to supply energy for contraction when the oxygen supply becomes limiting; in the foetus and new-born to tide over a period of hypoxia; in the intestinal wall to provide energy for absorption; and in the retina to supply energy by a mechanism which does not interfere with the light-transparency of the tissue.

XII. The Crabtree Effect

The Crabtree effect—the inhibition of respiration by glycolysis—is a much less common phenomenon than the Pasteur effect. The inhibition is small (10-50%) and occurs only in a few types of cells which possess a high glycolytic capacity, such as ascites tumour cells and other neoplastic tissues, renal medulla, leucocytes and cartilage (for review see[133]). The explanation of the Crabtree effect is analogous to that of the Pasteur effect, though allosteric mechanisms are not involved. But the same substances which are responsible for the allosteric effects on phosphofructokinase—ATP, ADP and P_i—are key factors in the Crabtree effect. The rate of respiration depends on the availability of ADP and P_i and on the ratio $[ATP]/[ADP][P_i]$. When glycolysis keeps [ADP] and $[P_i]$ low and the phosphorylation state of the adenine nucleotides high the rate of respiration is bound to be low. Thus while the Pasteur effect is caused by allosteric inhibition of phosphofructokinase, the Crabtree effect is caused by removal, through glycolysis, of the substrates of oxidative phosphorylation.[134,135] Another way of expressing this situation is to say that glycolysis and the respiratory chain both compete for ADP and P_i, and when glycolysis wins the result is a Crabtree effect.

XIII. Meyerhof's Experiments on the Pasteur Effect in Muscle

As stated at the beginning of this essay, early experiments by Meyerhof were of great historical importance in that they established the fact that the non-appearance of the products of fermentation in the presence of oxygen could not be caused by their complete oxidation. But it must also be said that Meyerhof's experiments on frog muscle (which purported to demonstrate the removal of lactate and an approximately equivalent formation of glycogen), proved irreproducible in the hands of other investigators (for example Eggleton and Evans[136]) and the concept of a resynthesis of glycogen from lactate in muscle was subsequently abandoned. One of the reasons for this was probably the fact that the key enzymes necessary for the resynthesis, especially those responsible for the conversion of pyruvate into phosphopyruvate, are very weak if not totally lacking in muscle.[137]

Recently, Bendall and Taylor[138] have reinvestigated Meyerhof's claim of the occurrence of glycogen synthesis, using frog muscle and psoas muscle of the rabbit. They come to the conclusion "the present results appear to vindicate Meyerhof's original hypothesis in its two most important aspects: first that the total amount of lactate disappearing from frog muscle during aerobic recovery from a tetanus or a period of anoxia is 5-6 times the amount that is oxidised; secondly that the missing lactate reappears in muscle in the form of glycogen". These findings are puzzling and contrary to those of other investigators.[139] In the isolated perfused hindquarter of the rat, a preparation which permits quantitative experiments to be carried out under near-physiological conditions, Houghton[140] found no uptake of added lactate by muscle during rest or during recovery after severe exercise, even with a lactate concentration in the perfusion medium of 5 mM.

The experiments of Bendall and Taylor[138] were carried out, it should be emphasized, under highly abnormal conditions. In order to demonstrate glycogen formation in frog muscles, the tissue had to be first incubated anaerobically for very long periods (16-20 hours) in order to cause an accumulation of lactate. This was followed by 4 to 6 hour periods of experimental incubation, during which a synthesis of glycogen was reported to occur. In order to detect glycogen formation in the rabbit muscle, pieces of the tissue were incubated 4 to 6 h with 44 mM lactate—the concentration of blood lactate *in vivo* never exceeds 12 mM. Furthermore, compared with the metabolic rates which occur *in vivo*, the rates observed in these experiments were exceedingly low so that even if measurable they would be negligible in relation to overall events *in vivo*.

XIV. Glycolysis and Lipogenesis

A special feature of glycolysis in liver, adipose tissue and mammary gland is its involvement in the conversion of carbohydrate into fat. In fact in these tissues the participation in the synthesis of fat is the main physiological function of glycolysis. For this reason glycolysis in these organs has special characteristics: in liver and adipose tissue it is dependent upon the dietary state of the organism; in the mammary gland it depends, of course, on the functional state of the organ.

Lipogenesis from carbohydrate involves glycolysis to the stage of pyruvate, the conversion of pyruvate to acetyl CoA, the synthesis of fatty acids from acetyl CoA and the esterification of the fatty acid formed. As the steps from glucose to pyruvate are those of glycolysis they can proceed anaerobically provided that the NADH formed is removed by the reduction of pyruvate to lactate. But the dehydrogenation of pyruvate and the subsequent stages depend, as experiments demonstrate, on the availability of oxygen, especially in the liver. The reason is probably as follows: the synthesis of fatty acids from acetyl CoA requires two molecules of $NADPH_2$ and two molecules of ATP per molecule of acetyl CoA. Although glycolysis to the stage of pyruvate supplies the required amounts of ATP and the required reducing equivalents in the form of mitochondrial $NADH_2$, extra energy is needed for the transfer of the reducing equivalents to the cytoplasm, and for the establishment of the necessary redox state of the cytoplasmic NADP-couple.

The concept that the anaerobic glycolysis of liver and adipose tissue represent a component of lipogenesis is supported by the facts that liver glycolysis is minimal in starvation and high after carbohydrate feeding, and that the rate of lactate production in the liver under anaerobic conditions is approximately equivalent to the rate of triglyceride synthesis.[141] In adipose tissue the rate of anaerobic glycolysis is likewise dependent upon the nutritional state,[142] but whereas in the liver the anaerobic disappearance of carbohydrate is equivalent to the formation of lactate, in adipose tissue less lactate is formed than carbohydrate removed.[143] Thus it appears that some fat can be synthesized in this tissue anaerobically.

The participation of glycolysis in lipogenesis implies that at the sites of lipogenesis the regulatory mechanisms which normally inhibit aerobic glycolysis do not operate. How this is achieved has already been discussed in Section VIII.

XV. Aerobic and Anaerobic Removal of Glucose

In those cells—and these are the majority—where the provision of energy by lactate formation and by hexose oxidation are the only major processes by which glucose or fructose is consumed, the aerobic rate of hexose removal, because of the Pasteur effect, is always lower, usually very much lower, than the anaerobic rate. In a few tissues however hexoses undergo other major reactions. Liver can form large amounts of glycogen from glucose and other hexoses, and kidney converts fructose into glucose. It is true that other tissues can also form some glycogen, but in relation to the energy yielding reactions their rate of glycogen deposition is usually low.

Since glycogen deposition and glucose formation from other hexoses depend on the supply of ATP they are very much more rapid in the presence of oxygen, and the aerobic rate of hexose removal can therefore be much greater than the anaerobic rate in liver and kidney cortex.

A special situation appears to obtain in the intestinal mucosa where not only the removal of glucose or fructose but also the formation of lactate can be greater under aerobic than under anaerobic conditions. Lohmann, Graetz and Langen,[144] who observed this phenomenon in isolated mucosa, described it as a "negative Pasteur effect". It is absent from the mucosa of newborn rats and mice and appears within 2 to 4 weeks after birth.

XVI. General Comments on the Relations Between Respiration and Fermentation

Reference has already been made to the fact (Section V) that in most tissues the total energy supply, in terms of ATP production, is much lower anaerobically than aerobically. Since ATP production is regulated by ATP consumption it follows that aerobically a greater volume of energy consuming processes takes place. This is not surprising because anaerobiosis is liable to cause a major derangement in the chemical environment within the cell which is unfavourable for some metabolic processes though still compatible with the maintenance of life. Thus in the anaerobic liver the concentrations of ATP, NAD, NADP, acetyl-CoA, pyruvate, oxaloacetate and α-oxoglutarate fall, while those of AMP, P_i, the phosphorylated hexoses, α-glycerophosphate, lactate, malate, alanine and aspartate rise.[145,146] Most of the changes are connected with the shift of the redox state of the pyridine nucleotide couples and their substrates in the direction of reduction.

So at best anaerobic fermentation may be expected to replace the aerobic ATP supply, provided that the rate of fermentation is high enough, but there are other features of respiration which fermentation itself cannot replace: it cannot maintain the internal environment of the cell in respect to the redox state of the pyridine nucleotides and other redox catalysts. For this the great majority of cells require a continuous supply of oxygen.

Microorganisms which can grow anaerobically in simple media have evolved special mechanisms for the maintenance of the redox state. The energy supply by lactic acid or alcoholic fermentation is neutral with respect to changes of the redox balance, since oxidations and reductions are equal. However, when cells grow on a simple medium with glucose as the sole source of carbon they have to synthesize many cell constituents by reactions which involve an excess of oxidations over reductions. Only a small minority of syntheses involve an excess of reductions while some biosyntheses are balanced with respect to oxidations and reductions. A survey of the redox balance of some of the quantitatively more important biosyntheses is given in Table 6.

TABLE 6

Examples of changes in the redox balance during biosynthesis from glucose in microorganisms. The synthesis of a few amino acids (aspartate, alanine) is neutral in respect to the redox balance. The majority of biosyntheses of amino acids involve a surplus of oxidations over reductions and therefore supply H atoms which need to be disposed of. In most cases NAD or NADP is the primary hydrogen acceptor. For details of pathways see Dagley and Nicholson.[147]

Substance synthesized	Number of H atoms arising	Intermediate stages of synthesis
Glutamate	6	pyruvate $\rightarrow$ oxaloacetate $\rightarrow$ citrate
Proline	2	glutamate
Arginine	2	glutamate
Serine	2	phosphoglycerate
Glycine	6	serine
Ribose	4	glucose 6-P $\rightarrow$ 6-phosphogluconate
Cystine	6	serine $\rightarrow$ cysteine
Lysine	10	oxoglutarate $\rightarrow$ aminoadipate $\rightarrow$ saccharopine

Microorganisms which grow anaerobically in simple media must balance all oxido-reductions and they do this by providing special hydrogen acceptors which replace O_2. Yeast and many bacteria can form succinate from glucose. This arises by CO_2 fixation and hydrogenation, the balance being:

$$\tfrac{1}{2}C_6H_{12}O_6 + CO_2 + 2H \rightarrow COOH . CH_2 . CH_2COOH$$

Many Enterobacteria form ethanol, H_2 and CO_2 as end-products and, as in the formation of succinate, the overall balance of this process is a hydrogenation:

$$\tfrac{1}{2}C_6H_{12}O_6 + 2H \rightarrow CH_3 . CH_2OH + CO_2 + H_2$$

The significance of these "mixed" fermentations has in the past been puzzling. The present considerations indicate that they are a matter of design. The various products can make a contribution to the regulation of the redox balance in the cell, especially of the pyridine nucleotide couples when biosyntheses upset the normal redox balance by the prevalence of oxidative steps.

XVII. Concluding Remarks

It is fair to state that the nature of the Pasteur effect is now known in principle though there are details still to be clarified. The early assumptions— first that the products of fermentation do not appear aerobically because they are burnt, or that the products are resynthesized to carbohydrate at the expense of the energy provided by respiration—have proved erroneous. It is now clear that the allosteric properties of phosphofructokinase can account for most aspects of the Pasteur effect. Retrospectively, it may be said that a fundamental solution of the problem posed by the Pasteur effect was not possible in the 1920's when Warburg and Meyerhof clearly formulated the chemical and energetic aspects. At that time knowledge of the pathway of glycolysis, of general enzymology and of the nature of regulatory mechanisms was far too scanty. Before the problem could be effectively tackled, the enzymes of glycolysis had to be identified, purified and studied in isolation. The key role of ATP in energy transformations had to be discovered. It had to be recognized that some enzymes are not only powerful and specific catalysts, but also possess regulatory sites, in addition to their catalytic sites. And the concept of allostery had to be formulated and developed.

During his long debate with Liebig about the relations between fermentation and the life of the yeast cell, Pasteur[148] in 1876 offered the comment: "The life-blood of science is the search for answers to successive questions of ever increasing subtlety and of ever increasing closeness to the very nature of the matter". The development of the knowledge of the Pasteur effect bears out the wisdom of this statement.

REFERENCES

1. Pasteur, L. (1861). **Expériences** et vues nouvelles sur la nature des fermentations. *C.R. Acad. Sci.* **52**, 1260-1264.
2. Pasteur, L. (1875). Nouvelles observations sur la nature de la fermentation alcoolique. *C. R. Acad. Sci.* **80**, 452-457.
3. Fletcher, W. M. & Hopkins, F. G. (1907). Lactic acid in amphibian muscle. *J. Physiol. (London)*, **35**, 247-309.
4. Fletcher, W. M. & Hopkins, F. G. (1917). The respiratory process in muscle and the nature of muscular motion. *Proc. Roy. Soc. Ser. B.* **89**, 444-467.
5. Meyerhof, O. (1930). *Die chemischen Vorgange im Muskel und ihr Zusammenhang mit Arbeitsleistung und Wärmebildung,* Springer Verlag, Berlin.
6. Meyerhof, O. (1920). Die Energieumwandlungen im Muskel. I. Über die Beziehungen der Milchsäure zur Wärmebildung und Arbeitsleistung des Muskels in der Anaerobiose. *Pflüg. Arch. ges. Physiol.* **182**, 232-283.

7. Meyerhof, O. (1921). Über die Milchsaurebildung in der zerschnittenen Muskulatur. *Pflüg. Arch. ges Physiol.* **188**, 114-160.
8. Warburg, O. (1926). *Über den Stoffwechsel der Tumoren,* Springer Verlag, Berlin.
9. Goddard, D. R. & Meeuse, B. J. D. (1950). Respiration of higher plants. *Annu. Rev. Plant Physiol.* **1**, 207-232.
10. Warburg, O. (1926). Über die Wirkung von Blausäureäthylester (Äthylcarbylamin) auf die Pasteursche Reaktion. *Biochem. Z.* **172**, 432-441.
11. Burk, D. (1937). On the biochemical significance of the Pasteur Reaction and Meyerhof cycle in intermediate carbohydrate metabolism. *Occasional Publications of the American Association for the Advancement of Science* **4**, 121-161.
12. Burk, D. (1939). A colloquial consideration of the Pasteur and neo-Pasteur effects. *Cold Spring Harb. Symp. quant. Biol.* **7**, 420-459.
13. Lipmann, F. (1933). Über die oxydative Hemmbarkeit der Glykolyse und den Mechanismus de Pasteurschen Reaktion. *Biochem. Z.* **265**, 133-140.
14. Lipmann, F. (1934). Über die Hemmung der Mazerationsgärung durch Sauerstoff in Gegenwart positiver Oxydationssysteme. *Biochem. Z.* **268**, 205-213.
15. Krebs, H. A. & Veech, R. L. (1969). Equilibrium relations between pyridine nucleotides and adenine nucleotides and their roles in the regulation of metabolic processes. *Adv. Enzyme Regul.* **7**, 397-413.
16. Lynen, F. (1941). Über den aeroben Phosphatbedarf der Hefe. Ein Beitrag zur Kenntnis der Pasteurschen Reaktion. *Justus Liebigs Ann. Chem.* **546**, 120-141.
17. Johnson, M. (1941). The role of aerobic phosphorylation in the Pasteur effect. *Science,* **94**, 200-202.
18. Wu, R. & Racker, E. (1959). Limiting factors in glycolysis of ascites tumour cells. *J. Biol. Chem.* **234**, 1029-1035.
19. Magne, H., Mayer, A. & Plantefol, L. (1932). Action pharmacodynamique des phenols nitres. Un agent augmentant les oxydations cellulaires. Le dinitrophenol 1.2.4. (Thermol). *Ann. Physiol. et de Physiochemie Biologique,* **8**, 1-50.
20. Dodds, E. C. & Greville, G. D. (1934). Effect of dinitrophenol on tumour metabolism. *Lancet* **I**, 398-399.
21. Clowes, G. H. A. & Krahl, M. E. (1936). Studies on cell metabolism; on relations between molecular structures, chemical properties and biological activities of nitrophenols. *J. Gen. Physiol.* **20**, 145-171.
22. Clifton, C. A. (1940). On the possibility of preventing assimilation in respiring cells. *Enzymologia,* **4**, 246-253.
23. Loomis, W. F. & Lipmann, F. (1948). Reversible inhibition of the coupling between phosphorylation and oxidation. *J. Biol. Chem.* **173**, 807-808.
24. Engelhardt, V. A. & Sakov, N. E. (1943). The mechanism of the Pasteur effect. *Biokhimiya,* **8**, 9-36.
25. Dickens, F. (1951). The Pasteur Effect. In *The Enzymes* (Sumner, J. S. & Myrbäck, K., eds.) vol. 2, pp. 672-688, Academic Press, London and New York.
26. Lynen, F., Hartmann, G., Netter, K. F. & Schuegraf, A. (1959). Phosphate turnover and Pasteur effect. In *Ciba Foundation Symposium on Regulation of Cell Metabolism,* (Wolstenholme, G. E. W. & O'Connor, C. M., eds.) pp. 256-273, J. H. Churchill, London.

27. Lonberg-Holm, K. K. (1959). A direct study of intracellular glycolysis in Ehrlich's ascites tumour. *Biochim. Biophys. Acta*, **35**, 464-472.

28. Park, C. R., Morgan, H. E., Henderson, J. J., Regen, D. M., Cadenas, E. & Post, R. L. (1961). The regulation of glucose uptake in muscle as studied in the perfused rat heart. *Recent Progr. Hormone Res.* **17**, 493-538.

29. Newsholme, E. A. & Randle, P. J. (1961). Regulation of glucose uptake by muscle. 5. Effects of anoxia, insulin, adrenaline and prolonged starving on concentrations of hexose phosphates in isolated rat diaphragm and perfused isolated rat heart. *Biochem. J.* **80**, 655-662.

30. Iwakawa, Y. (1944). Studies on the Pasteur reaction in muscle. *J. Biochem. (Tokyo)*, **36**, 191-326.

31. Bücher, Th. (1959). Zelluläre Koordination Energie transformierender Stoffwechselketten. *Ang. Chem.* **71**, 744.

32. Mansour, T. E. & Mansour, J. M. Effects of serotonin (5-hydroxytryptamine) and adenosine 3,5-phosphate on phosphofructokinase from the liver fluke, *Fasciola hepatica. J. Biol. Chem.* **237**, 629-634.

33. Passonneau, J. V. & Lowry, O. H. (1962). Phosphofructokinase and the Pasteur effect. *Biochem. Biophys. Res. Commun.* **7**, 10-15.

34. Passonneau, J. V. & Lowry, O. H. (1964). The role of phosphofructokinase in metabolic regulation. *Adv. Enzyme Regul.* **2**, 265-274.

35. Wu, R. (1964). Control of glycolysis by phosphofructokinase in slices of rat liver, Novikoff hepatoma and adenocarcinomas. *Biochem. Biophys. Res. Commun.* **14**, 79-85.

36. Viñuela, E., Salas, M. & Sols, A. (1963). Endproduct inhibition of yeast phosphofructokinase by ATP. *Biochem. Biophys. Res. Commun.* **12**, 140-145.

37. Garland, P. B., Randle, P. J. & Newsholme, E. A. (1963). Citrate as an intermediary in the inhibition of phosphofructokinase in rat heart muscle by fatty acids, ketone bodies, pyruvate, diabetes and starvation. *Nature (London)*, **200**, 169-170.

38. Parmeggiani, A. & Bowman, R. H. (1963). Regulation of phosphofructokinase activity by citrate in normal and diabetic muscle. *Biochem. Biophys. Res. Commun.* **12**, 268-273.

39. Mansour, T. E., Wakid, N. & Sprouse, H. M. (1966). Studies on heart phosphofructokinase: purification, crystallisation and properties of sheep heart phosphofructokinase. *J. Biol. Chem.* **241**, 1512-1521.

40. Ahlfors, C. E. & Mansour, T. E. (1969). Studies on heart phosphofructokinase: desensitization of the enzyme to adenosine triphosphate inhibition. *J. Biol. Chem.* **244**, 1247-1251.

41. Paetkau, V. & Lardy, H. A. (1967). Phosphofructokinase. Correlation of physical and enzymatic properties. *J. Biol. Chem.* **242**, 2035-2042.

42. Hulme, E. C. & Tipton, K. F. (1971). The dependence of phosphofructokinase kinetics upon protein concentration. *FEBS Lett.* **12**, 197-200.

43. Ho, W. & Anderson, J. W. (1971). Phosphofructokinase in rat jejunal mucosa; subcellular distribution, isolation and characterisation. *Biochim. Biophys. Acta*, **227**, 354-363.

44. Ibsen, K. H. & Schiller, K. W. (1971). Control of glycolysis and respiration in substrate depleted Ehrlich ascites tumour cells. *Archs Biochem. Biophys.* Vol. 1, **143**, 187-203.

45. Mansour, T. E. (1970). Kinetic and physical properties of phospho-fructokinase. *Adv. Enzyme Regulation*, **8**, 37-52.
46. Krzanowski, J. & Matschinsky, F. M. (1969). Regulation of phospho-fructokinase by phosphocreatine and phosphorylated glycolytic inter-mediates. *Biochem. Biophys. Res. Commun.* **34**, 816-823.
47. Mansour, T. E. (1965). Studies on heart phosphofructokinase: active and inactive forms of the enzyme. *J. Biol. Chem.* **240**, 2165-2172.
48. Weil-Malherbe, H. & Bone, A. D. (1951). The hexokinase activity of rat-brain extracts. *Biochem. J.* **49**, 339-347.
49. Crane, R. K. & Sols, A. (1954). The non-competitive inhibition of brain hexokinase by glucose 6-phosphate and related compounds. *J. Biol. Chem.* **210**, 597-606.
50. Grossbard, L. & Schimke, R. T. (1966). Multiple hexokinases of rat tissues. *J. Biol. Chem.* **241**, 3546-3560.
51. England, P. J. & Randle, P. J. (1967). Effectors of rat-heart hexokinase and the control of rates of glucose phosphorylation in the perfused rat heart. *Biochem. J.* **105**, 907-920.
52. Randle, P. J., Denton, R. M. & England, P. J. (1968). Citrate as a metabolic regulator in muscle and adipose tissue. In *The Metabolic Roles of Citrate*, (Goodwin, T. W., ed.), *Biochemical Society Symposium*, **27**, pp. 87-1034.
53. Newsholme, E. A. & Gevers, W. (1967). Control of glycolysis and gluconeogenesis in liver and kidney cortex. *Vitamins and Hormones*, **25**, 1-87.
54. Williamson, J. R., Jones, E. A. & Azzone, G. F. (1964). Metabolic control in perfused rat heart during fluoracetate poisoning. *Biochem. Biophys. Res. Commun.* **17**, 696-702.
55. Williamson, J. R. (1965). Glycolytic control mechanisms. *J. Biol. Chem.* **240**, 2308-2321.
56. Bassham, J. A. & Krause, G. H. (1969). Free energy changes and metabolic regulation in steady-state photosynthetic carbon reduction. *Biochem. Biophys. Acta*, **189**, 207-221.
57. Gumaa, K. A., McLean, P. & Greenbaum, A. L. (1971). Compartmentation in relation to metabolic control in liver. In *Essays in Biochemistry*, (Campbell, P. N. & Dickens, F, eds.), Vol. 7, pp. 39-86, Academic Press, London and New York.
58. Fitch, W. M. & Chaikoff, I. L. (1960). Extent and pattern of adaptation of enzyme activities in livers of normal rats fed diets high in glucose ˛and fructose. *J. Biol. Chem.* **235**, 554-557.
59. Loebel, R. O. (1925). Beiträge zur Atmung und Glykolyse tierischer Gewebe. *Biochem. Z.* **161**, 219-239.
60. Dickens, F. & Greville, G. D. (1932). The anaerobic conversion of fructose into lactic acid by tumour and adult normal tissues. *Biochem. J.* **26**, 1546-1556.
61. Dickens, F. & Greville, G. D. (1933). Respiration in fructose and in sugar-free media. *Biochem. J.* **27**, 832-841.
62. Sols, A. (1968). Phosphorylation and glycolysis. In *Carbohydrate Meta-bolism and its Disorders*, (Dickens, F., Randle, P. J. & Whelan, W. J., eds.), Vol. 1, pp. 53-87, Academic Press, London and New York.
63. Walker, D. G. (1966). The nature and function of hexokinase in animal

tissues. In *Essays in Biochemistry*, (Campbell, P. N. & Greville, G. D., eds.), 2, pp. 33-67, Academic Press, London and New York.

64. Kjerulf-Jensen, K. (1942). The hexosemonophosphoric acids formed within the intestinal mucosa during absorption of fructose, glucose and galactose. *Acta Physiol. Scand.* **4**, 225-248 and The phosphate esters formed in the liver tissue of rats and rabbits during assimilation of hexose and glycerol. *Acta Physiol. Scand.* **4**, 249-258.

65. Hers, H. G. (1957). Le métabolisme du fructose, Editions Arscia, Bruxelles.

66. Heinz, F. (1968). Messung der Enzymaktivitäten in der Dünndarmmucosa der Ratte. *Hoppe-Seyler's Z. physiol. Chem.* **349**, 339-344.

67. Burch, H. B., Max, P., Chyu, K. & Lowry, O. H. (1969). Metabolic intermediates in liver of rats given large amounts of fructose or dihydroxyacetone. *Biochem. Biophys. Res. Commun.* **34**, 619-626.

68. Adelman, R. C. Ballard, F. J. & Weinhouse, S. (1967). Purification and properties of rat liver fructokinase. *J. Biol. Chem.* **242**, 3360-3365.

69. Woods, H. F., Eggleston, L. V. & Krebs, H. A. (1970). The cause of hepatic accumulation of fructose 1-phosphate on fructose loading. *Biochem. J.* **119**, 501-510.

70. Frandsen, E. K. & Grunnet, N. (1971). Kinetic properties of triokinase from rat liver. *Eur. J. Biochem.* **23**, 588-592.

71. Heinz, F. (1968). Enzyme des Fructosestoffwechsels. Änderungen von Enzymaktivitäten in Leber und Niere der Ratte bei fructose-und glucosereicher Ernährung. *Hoppe-Seyler's Z. Physiol. Chem.* **349**, 399-404.

72. Warburg, O. & Hiepler, E. (1952). Versuche mit Ascites-Tumorzellen. *Z. Naturforsch*, **7b**, 193-194.

73. Krebs, H. A. (1964). Gluconeogenesis. *Proc. Roy. Soc. Ser. B.* **159**, 545-564.

74. Hems, R., Ross, B. D., Berry, M. N. & Krebs, H. A. (1966). Gluconeogenesis in the perfused rat liver. *Biochem. J.* **101**, 284-292.

75. Margaria, R., Edwards, H. T. & Dill, D. B. (1933). The possible mechanism of contracting and paying the oxygen debt and the role of lactic acid in muscular contraction. *Amer. J. Physiol.* **106**, 689-715.

76. Margaria, R. (1967). Aerobic and anaerobic energy sources in muscular exercise. In *Exercise in Altitude,* Excerpta Medica Foundation, New York.

77. Dickens, F. & Weil-Malherbe, H. (1941). Metabolism of intestinal mucous membrane. *Biochem. J.* **35**, 7-15.

78. Wiseman, G. (1964). *Absorption from the Intestine*, Academic Press, London and New York.

79. Wilson, T. H. (1962). *Intestinal Absorption*, W. B. Saunders Co., Philadelphia and London.

80. Kiyasu, J. Y. & Chaikoff, I. L. (1957). On the manner of transport of absorbed fructose. *J. Biol. Chem.* **224**, 935-939.

81. Srivastava, L. M. & Hübscher, G. (1966). Glucose metabolism in the mucosa of the small intestine. Glycolysis in subcellular preparations from the cat and rat. *Biochem. J.* **100**, 458-466.

82. Stifel, F. B., Rosensweig, N. S., Zakim, D. & Herman, R. H. (1968). Dietary regulation of glycolytic enzymes. I. Adaptive changes in rat jejunum. *Biochim. Biophys. Acta*, **170**, 221-227.

83. Rosensweig, N. S., Stifel, F. B., Herman, R. H. & Zakim, D. (1968). The dietary regulation of the glycolytic enzymes. II. Adaptive changes in human jejunum. *Biochim. Biophys. Acta*, **170**, 228-234.
84. Dickens, F. & Weil-Malherbe, H. (1936). A note on the metabolism of medulla of kidney. *Biochem. J.* **30**, 659-660.
85. Kean, E. L., Adams, P. H., Winters, R. W. & Davies, R. E. (1961). Energy metabolism of the renal medulla. *Biochim. Biophys. Acta*, **54**, 474-478.
86. Kean, E. L., Adams, P. H., Conrad Davies, H., Winters, R. W. & Davies, R. E. (1962). Oxygen consumption and respiratory pigments of mitochondria of the inner medulla of the dog kidney. *Biochim Biophys. Acta*, **64**, 503-507.
87. Ruiz-Guinazu, A., Pehling, G., Rumrich, G. & Ullrich, K. J. (1961). Glucose and lactic acid concentration at the peak of the vascular counterflow system in the renal medulla. *Pflüg. Arch. ges Physiol.* **274**, 311-317.
88. Ullrich, K. J., Kramer, K. & Boylan, J. W. (1961). Present knowledge of the counter-current system in the mammalian kidney. *Progress in Cardiovascular Diseases*, **3**, 395-431.
89. Ullrich, K. J., Pehling, G. & Stöckle, H. (1961). Hämoglobinkonzentration, Erythrocytenzahl und hämatokrit im vasa recta Blut. *Pflüg. Arch. ges. Physiol.* **273**, 573-578.
90. Harris, J. E. (1941). The influence of the metabolism of human erythrocytes on their potassium content. *J. Biol. Chem.* **141**, 579-595.
91. Rodnan, G. P., Ebaugh, F. G. Jr., Fox, M. R. S. (1957). The life span of the red blood cell and the red blood cell volume in the chicken, pigeon and duck as estimated by the use of $Na_2^{51}CrO_4$, with observations on red cell turnover rate in the mammal, bird and reptile. *Blood (Baltimore)*, **12**, 355-366.
92. Benesch, R., Benesch, R. E. & Yu, C. I. (1968). Reciprocal binding of oxygen and diphosphoglycerate by human haemoglobin. *Proc. Nat. Acad. Sci. U.S.* **59**, 526-532.
93. Rapoport, S. (1968). The regulation of glycolysis in mammalian erythrocytes. In *Essays in Biochemistry*, (Campbell, P. N. & Greville, G. D., Eds.), **4**, 68-103, Academic Press, London and New York.
94. Eaton, J. W., Brewer, G. J., Schultz, J. S. & Sing, C. F. (1970). Variation in 2,3-diphosphoglycerate and ATP levels in human erythrocytes and effects on oxygen transport. In *Red Cell Metabolism and Function*,(Brewer, G. J., ed.) pp. 21-38, Plenum Press, New York and London.
95. Gerlach, E., Duhm, J. & Deuticke, B. (1970). Metabolism of 2,3-Diphosphoglycerate in red blood cells under various experimental conditions. In *Red Cell Metabolism and Function*, (Brewer, G. J., ed.) pp. 155-174, Plenum Press, New York and London.
96. Benesch, R. & Benesch, R. E. (1969). Intracellular organic phosphates as regulators of oxygen release by haemoglobin. *Nature (London)*, **221**, 618-622.
97. Bartlett, G. R. (1970). Patterns of phosphate compounds in red blood cells of man and animals. In *Red Cell Metabolism and Function*, (Brewer, G. J., ed.) pp. 245-256, Plenum Press, New York and London.
98. Rapoport, S. (1940). Phytic acid in avian erythrocytes. *J. Biol. Chem.* **135**, 403-406.

99. Molinari, E. & Hoffmann-Ostenhof, O. (1968). Über ein Enzymsystem das myo-Inosit zu Phytinsäure phosphorylieren kann. *Hopp-Seyler's Z. physiol. Chem.* **349**, 1792-1799.

100. Dische, Z. (1941). Sur l'interdépendance des divers enzymes du système glycolytique et sur la régulation automatique de leur activité dans les cellules. I. Inhibition de la phosphorylation du glucose dans les hématies par les acides mono- et diphosphoglycérique; état de l'acide diphosphoglycérique et phosphorylation du glucose. *Travaux des Membres de la Société de Chimie Biologique*, **23**, 1140-1148.

101. Brewer, G. J. (1969). Erythrocyte metabolism and function: hexokinase inhibition by 2,3-diphosphoglycerate and interaction with ATP and Mg^{2+}. *Biochim. Biophys. Acta*, **192**, 157-161.

102. Shelley, H. J. (1961). Glycogen reserves and their changes at birth and in anoxia. *Brit. Med. Bull.* **17**, 137-143.

103. Dicker, S. E. & Shirley, D. G. (1971). Rates of oxygen consumption and of anaerobic glycolysis in renal cortex and medulla of adult and new-born rats and guinea pigs. *J. Physiol. (London)*, **212**, 235-243.

104. Ballard, F. J. (1971). The development of gluconeogenesis in rat liver. Controlling factors in the newborn. *Biochem. J.* **124**, 265-274.

105. Dawes, G. S. (1968). *Foetal and neonatal physiology*. Yearbook Medical Publishers Inc., Chicago.

106. Bullough, J. (1958). Protracted foetal and neonatal asphyxia. *Lancet*, I, 999-1000.

107. Mott, J. C. (1961). The ability of young mammals to withstand total oxygen lack. *Brit. Med. Bull.* **17**, 144-147.

108. Okamoto, Y. (1925). Über Anaerobiose von Tumorgewebe. *Biochem. Z.* **160**, 52-65.

109. Warburg, O., Gawehn, K., Geissler, A., Schröder, W., Gewitz, H. S. & Völker, W. (1958). Partielle Anaerobiose und Strahlenempfindlichkeit der Krebszellen. *Arch. Biochem. Biophys.* **78**, 573-586.

109a Weber, G. (1972). The molecular correlation concept: recent advances and implications. *Gann-Monograph* No. 13, published by Jap. Cancer Ass., Maruzen Co. Ltd., Tokyo.

110. Warburg, O., Posener, K. & Negelein, E. (1924). Über den Stoffwechsel der Carcinomzelle. *Biochem. Z.* **152**, 309-343.

111. Krebs, H. A. (1927). Über den Stoffwechsel der Netzhaut. *Biochem. Z.* **189**, 57-59.

112. Kubowitz, F. (1929). Stoffwechsel der Froschnetzhaut. *Biochem. Z.* **204**, 475-478.

113. Graymore, C. N. (1969). General aspects of the metabolism of the retina. In *The Eye*, (Dawson, H., ed.), vol. I. pp. 601-652. Academic Press, London and New York.

114. Fujita, A. (1928). Über den Stoffwechsel der Körperzellen. *Biochem. Z.* **197**, 175-188.

115. Murata, H. (1969). Retinal metabolism in experimental diabetes in rabbits. *Acta. Soc. Ophthal. Jap.* **67**, 1182-1191.

116. Nakashima, M. (1929). Stoffwechsel der Fischnetzhaut bei verschiedenen Temperaturen. *Biochem. Z.* **204**, 478-481.

117. Michaelson, I. C. (1954). *Retinal circulation in man and animals*. C. C. Thomas. Springfield, Illinois.

118. Hawkins, R. A. & Krebs, H. A. Unpublished observations.
119. Friedman, E. & Smith, T. R. (1965). Estimation of retinal blood flow in animals. *Invest. Ophthalmol.* **4**, 1122-1128.
120. Trokel, S. (1964). Measurement of ocular blood flow and volume by reflective densitometry. *Arch. Ophthal.* **71**, 88-92.
121. Krebs, H. A. & Hems, R. Unpublished observations on rabbit retina suspended in vitreous body liquified by homogenization. This value is considerably higher than that recorded by Murata[115] and given in Table 2.
122. Krebs, H. A. & Hems, R. Unpublished observations.
123. Cohen, A. I. (1963). The fine structure of the visual receptors of the pigeon. *Exp. Eye Res.* **2**, 88-97; and Vertebrate retinal cells and their organisation. *Biol. Rev.* **38**, 427-459.
124. Kühn. H. & Dreyer, W. J. (1972). Light dependent phosphorylation of rhodopsin by ATP. *FEBS Lett.* **20**, 1-6.
125. de Vincentiis, M. (1951). Ulteriori osservazioni sul contenuto di acido piruvico e acido lattico nel vitreo. *Boll. Soc. Ital. Biol. Sper.* **27**, 309-312.
126. Ready, D. V. & Kinsey, V. E. (1960). Composition of the vitreous humor in relation to that of plasma and aqueous humors. *Arch. Ophthal.* **63**, 715-723.
127. Mountford, G. (1958). *Portrait of a Wilderness,* p. 144. Hutchinson, London.
128. Hess, G. (1951). *The Bird: Its Life and Structure,* Herbert Jenkins, London.
129. Warburg, O. (1928). Stoffwechsel der Karzinomzelle. *Verhandl. deutsch. Kongr, Inn. Med.* **40**, 11-18.
130. Warburg, O. & Christian, W. (1943). Gäarungsfermente im Blutserum von Tumor-Ratten. *Biochem. Z.* **314**, 407-414.
131. Warburg, O., Gawehn, K. & Geissler, A. W. (1957). Manometrie der Körperzellen unter physiologischen Bedingungen. *Z. Naturforsch.* **12B**, 115-119.
132. Warburg, O, Gawehn, K. & Geissler, A. W. (1957). Über die Wirkung von Wasserstoffperoxyd auf Krebszellen und auf embryonale Zellen. *Z. Naturforsch.* **12B**, 393-396.
133. Aisenberg, A. C. (1961). *The Glycolysis and Respiration of Tumours,* Academic Press, New York and London.
134. Gatt, S. & Racker, E. (1959). Crabtree effect in reconstructed systems. *J. Biol. Chem.* **234**, 1015-1023.
135. Wu, R. & Racker, E. (1959). Limiting factors in glycolysis of ascites tumor cells. *J. Biol. Chem.* **234**, 1029-1035.
136. Eggleton, M. G. & Evans, C. L. (1930). Lactic acid content of blood after muscular contraction under experimental conditions. *J. Physiol. (London)*, **70**, 269-293.
137. Krebs, H. A. & Woodford, M. (1965). Fructose 1,6-diphosphatase in striated muscle. *Biochem. J.* **94**, 436-445.
138. Bendall, J. R. & Taylor, A. A. (1970). The Meyerhof quotient and the synthesis of glycogen from lactate in frog and rabbit muscle. *Biochem. J.* **118**, 887-893.
139. Koeppe, R. E., Inciardi, N. F., Warnock, L. G. & Wilson, W. E. (1964). Some aspects of the metabolism of D- and L-lactic acid-2-^{14}C by rat skeletal muscle *in vivo. J. Biol. Chem.* **239**, 3609-3612.

140. Houghton, C. R. S. (1971). Studies on the metabolism of exercise with special reference to perfused, isolated skeletal muscle. *D. Phil. Thesis, Oxford University.*

141. Woods, H. F. & Krebs, H. A. (1971). Lactate production in the perfused rat liver. *Biochem. J.* **125**, 129-139.

142. Krebs, H. A. & Jones, H. M. Unpublished observations.

143. Halperin, M. L. & Denton, R. M. (1969). Regulation of glycolysis and L-glycerol 3-phosphate concentration in rat epididymal adipose tissue *in vitro. Biochem. J.* **113**, 207-214.

144. Lohmann, K., Graetz, H. & Langen, P. (1966). The metabolism of the small intestine. In *Current Aspects of Biochemical Energetics,* (Kaplan, N. O. & Kennedy, E. P., eds), pp. 111-126, Academic Press, London and New York.

145. Brosnan, J. T., Krebs, H. A. & Williamson, D. H. (1970). Effects of ischaemia on metabolite concentrations in rat liver. *Biochem. J.* **117**, 91-96.

146. Hems, D. A. & Brosnan, J. T. (1970). Effects of ischaemia on content of metabolites in rat liver and kidney *in vivo. Biochem. J.* **120**, 105-111.

147. Dagley, S. & Nicholson, D. E. (1970). *An Introduction into Metabolic Pathways,* Blackwell Scientific Publications, Oxford and Edinburgh.

148. Pasteur, L. (1876). *Etudes sur bière. Ses maladies, causes qui les provoquent, procédé pour la rendre inaltérable, avec une théorie novelle de la fermentation.* Chap. VI, Sect. VI, p. 315. Gauthier-Villars, Paris. The original text, the beauty of which is difficult to render in translation, is as follows: La science vit de solutions successives données à des *pourquoi* de plus en plus subtils, de plus en plus rapprochés de l'essence même des phénomènes.

Teichoic Acids in Cell Walls and Membranes of Bacteria

JAMES BADDILEY

*Microbiological Chemistry Research Laboratory, School of Chemistry,
University of Newcastle upon Tyne, Newcastle upon Tyne NE1 7RU, England*

I. Introduction

A. DEFINITION

It is convenient to give a simple general name to a group of compounds that have something in common, whether this be a feature of structure, function or natural location. For this reason the name "teichoic acid" has been given to members of a group of polymers containing phosphate which have been isolated from walls and cell contents of Gram-positive bacteria. The first representatives were all polymers of either glycerol phosphate or ribitol phosphate in which repeating units were joined together through phosphodiester linkages. In addition, these polymers usually possessed glycosyl and D-alanine ester substituents.[1] During the course of time other phosphorylated polymers have been identified in bacterial walls and the original description is considered to be too restrictive. Consequently, the term teichoic acid is now intended to include all bacterial wall, membrane or capsular polymers containing glycerol phosphate or ribitol phosphate residues.[2] Even with this broader definition problems arise, as there are sugar 1-phosphate polymers or oligosaccharide 1-phosphate polymers in the walls of some bacteria, and these are thus not really teichoic acids; polymers of this kind nevertheless closely resemble the teichoic acids in their properties, and especially in their probable function and mechanism of biosynthesis, so they will be discussed where appropriate in this essay.

Most of the studies relating to structure, biosynthesis and function of teichoic acids have been carried out on representatives isolated from washed walls of Gram-positive bacteria, where they frequently occur in amounts between 30-50% of the dry weight of the wall. These are much easier to study than are the less plentiful membrane teichoic acids, although it is believed that the latter are biologically the more important. Because some Gram-positive bacteria and probably all Gram-negative bacteria do not possess wall teichoic acids the importance of these polymers to the life processes of bacteria has not always been appreciated. However, the lipopolysaccharides in the walls of Gram-negative bacteria show structural similarity to teichoic acids and probably play a similar role; moreover, the universal occurrence of membrane teichoic acids in the Gram-positive organisms indicates their vital importance in cell function. In addition to the importance of teichoic acids and other wall components to the bacterial cell itself, they are significant in the relationship between the host and infecting organism in disease, where the macromolecules in the outer regions of the cell frequently possess antigenic properties. In fact, teichoic acids are particularly effective antigens and in many cases have been identified as group- or type-specific substances.[3] Finally, many antibiotics including the penicillins and cephalosporins exert their characteristic effects on bacteria by interfering in various ways with the formation of the cell wall, and in this connection the close interrelationship that exists between the biosynthetic routes to all wall components confirms the conclusion that the wall polymers in general are of some importance in the action of antibiotics.

B. STRUCTURE OF BACTERIAL WALLS AND MEMBRANES

An understanding of the general nature of bacterial walls and membranes is essential to the discussion of one of their components. In most Gram-positive bacteria the cell contents are surrounded by a cytoplasmic membrane which is a delicate structure comprising protein, phospholipids of various kinds and smaller amounts of glycolipids and membrane teichoic acid. These components form a typical bilayer in which the protein is on the outside of both surfaces, and the phospholipids are arranged so that their charged ends associate with the protein and their hydrophobic ends are directed towards the inner region of the bilayer. This membrane is surrounded by a relatively rigid physically strong wall which is thought to be permeable to most small and medium-sized molecules. At least 50% of the weight of the wall is composed of a complex polymer usually called mucopeptide or peptidoglycan; the rest of the wall is either teichoic acid, an acidic polysaccharide containing uronic acid residues, or both of these. Peptidoglycans are two-dimensional polymers; extending in one direction are glycan chains containing an alternating sequence of N-acetylglucosamine and its 3-O-lactyl ether derivative (N-acetylmuramic acid), joined together β-glycosidically at the respective 4-positions; in certain cases O-acetyl groups are present in some of the muramic acid residues. These glycan chains are cross-linked, to varying extent in different organisms, by peptide chains. The nature of the peptide chains in *Staphylococcus aureus* is illustrated in Fig. 1; linkage is to the carboxyl of muramic acid, and cross-linkage is achieved through the pentaglycine "bridge" which is attached to the ϵ-amino group of lysine in a neighbouring peptide chain. Although the composition of glycan chains is remarkably constant among different bacteria, the peptides vary considerably in their composition and degree of cross-linkage. The teichoic acids and other wall polysaccharides are covalently attached to the peptidoglycan.

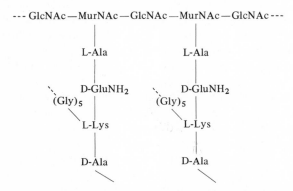

Fig. 1. Structure of peptidoglycan from *Staphylococcus aureus*.
GlcNAc = N-acetylglucosamine
MurNAc = N-acetylmuramic acid

The outer layers of Gram-negative bacteria are more complex. There is an inner membrane of similiar appearance to the cytoplasmic membrane in Gram-positive bacteria, a peptidoglycan layer outside it, and an outer membrane containing together with other components the lipopolysaccharide. Knowledge of the detailed morphology of the wall-membrane structure of Gram-negative bacteria is less complete than it is for the Gram-positive members, and will not be discussed in this essay. There is, however, considerable information available on peptidoglycans and lipopolysaccharides from these organisms, and the reader is referred to a book[4] on cell walls and membranes for references and other details.

II. Occurrence and Location

The teichoic acids were first detected as components of the walls of several Gram-positive bacteria, from which they could be extracted slowly by treatment with cold trichloroacetic acid solutions. By making use of one of the mechanical methods for the disruption of bacteria and recovery of their walls it can be shown that they are major wall components of many bacteria where they frequently represent between 20% and 50% of the dry weight of the wall. Their distribution is widespread, especially rich sources being the bacilli, lactobacilli, streptococci, staphylococci, micrococci and streptomyces. Their occurrence and structure can be of taxonomic importance; for example, all wild strains of *Staphylococcus aureus* contain ribitol teichoic acids in their walls. Similarly, the serological classification of certain bacteria,[3] e.g. lactobacilli, has been rationalized by a study of the chemical structure of their teichoic acids, which in many cases have been identified as the main characteristic serological components of the cell.

When trichloroacetic acid extracts of whole cells are compared with extracts of walls, it is found that the former always contain glycerol teichoic acids, even in the cases where the wall contains no teichoic acid. These compounds are clearly not situated in the wall, and at first they were called "intracellular teichoic acids". More information about their precise location has come from studies with protoplasts. When certain bacteria are treated with lysozyme in the presence of magnesium ions and sufficient sucrose or other substances to balance the osmotic pressure, wall-free protoplasts are formed which possess many of the properties of the whole cells. After such treatment the intracellular teichoic acid is mainly in the supernatant, and so it was concluded that it is located in the region between the wall and the cytoplasmic membrane.[5] On the other hand, in experiments carried out under slightly different conditions[6] much of the teichoic acid adhered to the protoplast surface. It appears that the magnesium ion concentration is largely responsible for the difference in results (unpublished work with Drs G. Pigott, J. Goundry and A. R. Archibald), and recent work suggests that a large part of this so-called "membrane teichoic acid" is in fact associated with the mesosomal particles. It is now known (see later) that membrane teichoic acids are covalently attached to lipids which presumably form a part of the membrane.

Although teichoic acids are characteristically wall and membrane components, polymers conforming to the chemical definition of teichoic acids occur in *Diplococcus pneumoniae* as capsular substances (ref. 7 *inter al.*). These are mainly ribitol derivatives with relatively complex structures.

The precise location of teichoic acids within the wall structure has received relatively little attention until recently. Attempts to locate them by electron microscopy may be criticized on the grounds of lack of the use of specific staining methods, whilst other attempts making use of the binding of specific antisera during the progressive dissolution of a wall with lysozyme make a number of questionable assumptions. In recent work with Drs J. Heckels and A. R. Archibald (unpublished), information on this matter has come from a study of teichoic acid-peptidoglycan linkages. The walls of *Staphylococcus lactis* 13 have been treated with an amidase preparation from a species of *Flavobacter* which removes almost completely the polypeptide component of peptidoglycan. The glycan chains are recovered in high yield, and it is found that nearly 40% of the glycan is covalently attached to teichoic acid; in this fraction where the two polymers are attached to each other, each glycan chain is attached to only one teichoic acid chain. All of the teichoic acid is attached to glycan and the average sizes of the two polymer chains are respectively 24 and 9 repeating units. The glycan chains that are not attached to teichoic acid also comprise 9 repeating disaccharide units. Not only does this suggest a considerable degree of order in wall structure and in the processes of wall synthesis, but it also has implications concerning the location of teichoic acids within the wall structure. If the glycan chains lie parallel to the surface of the wall, then the proportion of chains to which teichoic acid is attached would be considerably more than a single layer on the surface. The wall would have to be composed of several layers of glycan chains, 40% of which were attached to teichoic acid; it follows that the teichoic acid would permeate the wall structure fairly uniformly. Only if the glycan and teichoic acid chains lie perpendicular to the surface of the cell, presumably in a contracted conformation, could the teichoic acid occupy a discrete region either on the outer or inner surface of the wall.

III. Extraction Procedures

The teichoic acids were first extracted[1] from cells or walls of bacteria by treatment with cold dilute trichloroacetic acid solution followed by precipitation with organic solvents. The polymers are removed from the wall rather slowly, and extraction periods of 24-48 hours at 4°C, or even longer, are common. Although most of the work on the structure of these compounds has been carried out on material prepared in this way, it is likely that some of the phosphodiester linkages in the polymer chain have been hydrolysed during extraction. Reprecipitation from aqueous solution with acetone or ethanol, or ion-exchange chromatography, can be used to remove short-chain hydrolysis products, but the purified products are of limited value for the study of chain-length.

Dilute solutions of *N,N*-dimethylhydrazine at pH 7 in the presence of air or oxygen at 60°C for short periods dissolve the walls of many, but not all, bacteria.[8] The reaction probably involves free-radical attack on the peptidoglycan, and walls can be dissolved in some cases with the destruction of barely detectable amounts of sugar or amino acid residues. Teichoic acids are solubilized with no apparent degradation in this way, even when the wall itself does not dissolve. Small amounts of peptidoglycan components are usually present in the reprecipitated product, and the method has not been used extensively.

Several examples of the use of dilute alkali for the extraction of teichoic acids have been reported.[9-11] The conditions are gentle, 0·1 M-sodium hydroxide at 37°C usually effecting rapid extraction, and the procedure is likely to find extensive use. Although the phosphodiester linkages in polymers of glycerol phosphate and ribitol phosphate are stable under these conditions, the method cannot be used when alkali-labile linkages are present.

Enzymic methods have been used to degrade the peptidoglycan in walls, but in no case has it been possible to achieve this completely leaving intact the teichoic acid. Lysozyme is inhibited by many teichoic acids, and its action is thus incomplete in the hydrolytic removal of peptidoglycan from teichoic acid. One example has been reported of autolytic enzymes associated with a wall which bring about the almost complete removal of the peptidoglycan from the teichoic acid,[12] but this approach is not of general use.

IV. Structure

A. RIBITOL PHOSPHATE DERIVATIVES

The first teichoic acids to be subjected to structural investigation were all polymers of glycerol phosphate or ribitol phosphate in which repeating units are joined together by phosphodiester linkages.[1] The ribitol teichoic acids of this type are confined exclusively to bacterial walls, and the most extensively studied are those from strains of *Staphylococcus aureus*.[13,14] In all of those examined the phosphodiester linkages are between positions 1 and 5 on ribitol, and an *N*-acetylglucosaminyl residue is present at the D-4 position on each ribitol residue. The glycosidic linkages may be α or β, but in the majority of strains both types of linkage are present.[15] Most of the ribitol residues also possess a D-alanine ester residue, and the extracted polymer has an average chain comprising 6-10 repeating units. A structure is given in Fig. 2.

The methods used for characterizing structures such as that shown in Fig. 2

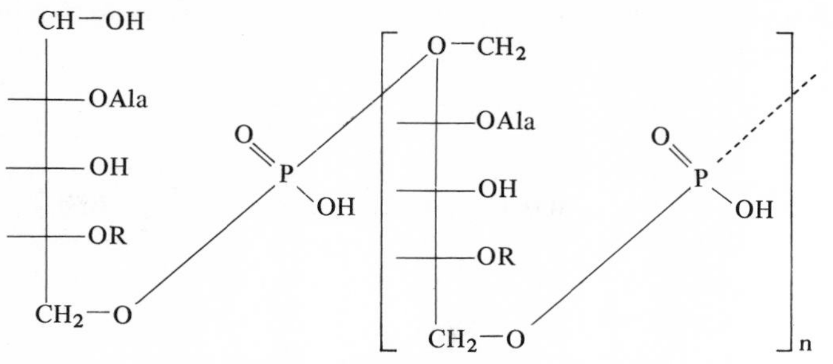

R = α or β-N-acetylglucosaminyl
Ala = D-alanyl

Fig. 2. Ribitol teichoic acid from the wall of *Staphylococcus aureus.*

have been discussed elsewhere[16] and will not be described here in detail; they include partial hydrolysis of polymers with acid or alkali to give repeating units, their isomers and further hydrolysis products which have been characterized by chemical degradation, enzymic methods and chromatography. Different approaches were necessary however in determining the exact location of the alanine ester residues, and in deciding whether each polymer molecule contains both α and β glycosyl substituents or whether the preparations are mixtures of polymers in which the individual molecules possess linkages of only one type. Although chemical methods[16] had indicated that the alanine ester residues must occupy hydroxyl groups at positions D-2 or 3 on ribitol, a decision between these alternatives was not possible at the time largely because the acidic conditions employed in the early work for extraction of teichoic acids would have caused ester migration between adjacent hydroxyl groups on the ribitol, and so the isolated polymer would have been an artificial equilibrium mixture of isomers. However, by using a mutant of the organism which has lost the ability to introduce glycosyl substituents onto ribitol it can be shown by periodate oxidation that the alanyl substituents occupy position D-2 on ribitol.[17] The question of the configuration of the glycosidic linkages in individual molecules has been settled by a serological method.[18] Precipitation reactions using antisera with specificity for *N*-acetylglucosaminyl residues having an established configuration indicate that samples of polymer are mixtures of molecules in which all of the glycosidic linkages have the same configuration.

Similar ribitol teichoic acids are found in the walls of bacillia and lactobacilli. In a strain of *Bacillus subtilis* the sugar substituents are β-glucopyranosyl,[19] whereas in *Lactobacillus arabinosus* 17-5 they are α-glucopyranosyl.[20] In the lactobacillus, additional α-glucopyranosyl substituents occur at position 3 on some of the ribitol residues.

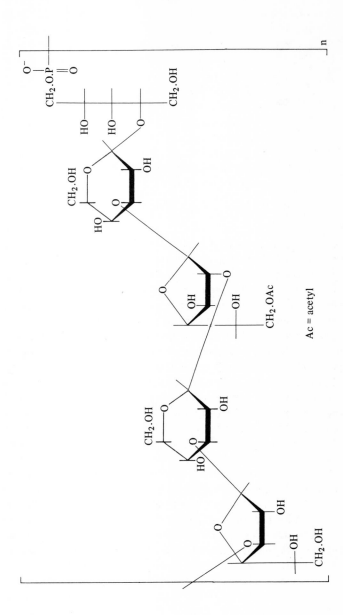

Fig. 3. Capsular substance from *Diplococcus pneumoniae* type 34.

Some bacterial cells possess, in addition to a wall, a relatively thick gelatinous coat or capsule. The capsular materials may be type-specific and the serological behaviour of an organism is often determined to a considerable extent by the nature of its capsular substances. For example, in *Diplococcus pneumoniae* the capsular substances are responsible for the behaviour of the organisms towards specific antisera, and type specificity has been correlated with the chemical structure of these capsular materials. A large number of pneumococcal types are known and their type-specific substances, all of which are carbohydrates, range from relatively simple polysaccharides through more complex representatives, some of which are teichoic acids. Those capsular substances that are ribitol teichoic acids differ from the polymers of ribitol phosphate described above in that the phosphodiester linkages are between ribitol and a hydroxyl on a sugar residue of a neighbouring unit; thus the sugar residues form a part of the polymer chain. In those examples where the detailed structure has been elucidated the carbohydrate components are oligosaccharides, frequently containing both pyranosyl and furanosyl residues and *N*-acetylamino sugars. The structure of the type-specific substance from the type 34 organism is given in Fig. 3; this compound contains in its repeating unit two galactofuranosyl, one galactopyranosyl, one glucopyranosyl, one acetyl and one ribitol phosphate.[21] Other even more complex ribitol teichoic acids occur amongst the pneumococcal capsular substances, and all apparently conform to the pattern in which complex oligosaccharides attached to ribitol phosphate constitute the main polymer chain (ref.[22] *inter al.*). Although these compounds are not wall components, their inclusion with the teichoic acids is justified on structural grounds. They differ however from the more conventional ribitol teichoic acids in lacking alanine ester residues. In fact, all pneumococci also possess ribitol teichoic acid in their walls, and this is believed to be identical in all of the large number of serological types of the organism. Its structure has not been established fully, but it is a derivative of ribitol phosphate with an oligosaccharide in the main chain; the sugar components are glucose, *N*-acetylgalactosamine and an *N*-acetyldiamino-dideoxy sugar in equimolar amounts.[23] A unique feature of this molecule is that, although it lacks alanine, it posseses a choline phosphate in its repeating structure. This teichoic acid has been identified as the serologically characteristic somatic antigen long-known as "C-substance". Actinomycetes also possess wall teichoic acids, including polymers of ribitol and glycerol phosphates. An interesting feature of some of these is the presence of succinyl ester substituents (work by Naumova and her colleagues, reviewed in Ref. 16).

B. GLYCEROL PHOSPHATE DERIVATIVES

Teichoic acids containing glycerol phosphate are more widespread than are those containing ribitol phosphate. The first examples to be subjected to

structural examination were polymers of glycerol phosphate in which phosphodiester linkages join the 1- and 3-positions of adjacent glycerol residues.[24] These
were membrane teichoic acids, and subsequent work[15,25,26] supports the
conclusion that all membrane teichoic acids are of this structural type. Many of
them do not have glycosyl substituents but a few are found on 2-hydroxyl
groups of occasional glycerol residues,[27] alanine ester residues occupy hydroxyl
groups on most of those glycerol residues that are not glycosylated; an example
is illustrated in Fig. 4. The ester residues are moderately stable towards gentle
acid treatment such as that required for the extraction of teichoic acids from
walls or intact cells, but are especially labile towards alkali and organic bases; the
presence of a hydroxyl or phosphate group adjacent to the alanyl ester
considerably enhances its reactivity towards bases.[28]

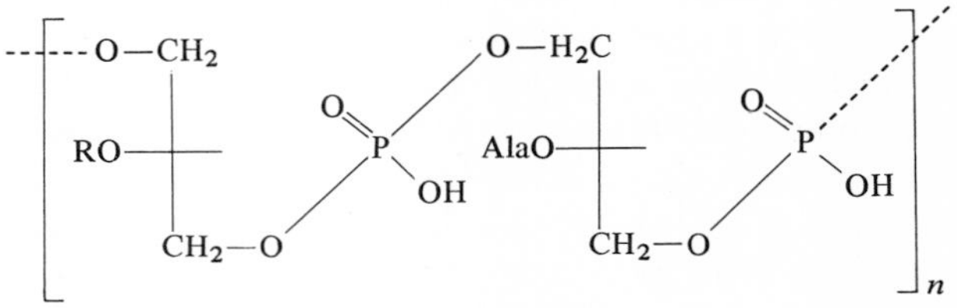

R = H or glycosyl
Ala = alanyl

Fig. 4. A typical glycerol teichoic acid.

An unusually high degree of glycosyl substitution is found in the membrane
teichoic acids of streptococci in serological group D. These are the group-specific
substances[29] in these organisms, and this high content of glycosyl residues is
probably responsible for their dominant serological behaviour. In *Streptococcus
faecalis* N.C.I.B. 8191 it was found[30] that each glycerol is substituted with
α-D-glucopyranosyl-(1 → 2)-α-D-glucopyranosyl (kojibiosyl) residues or with
kojitriosyl residues, and the alanyl ester substituents occupy unidentified sugar
hydroxyl groups (see Fig. 5). In a more recent study[31] a somewhat smaller
amount of glycosylation was found in material from the same organism; no
kojitriosyl residues were detected and not more than about 60% of the glycerol
residues were substituted with kojibiosyl. The reason for this different result has
not been established, but it illustrates a lack of reproducible consistency of
structure of teichoic acids that has been observed with other bacteria grown in
the laboratory over long periods where it has not been possible to control
accurately the composition of media and other growth conditions.

As with the ribitol teichoic acids, the commonly encountered sugar
substituents on polymers of glycerol phosphate are glucose and acetylamino

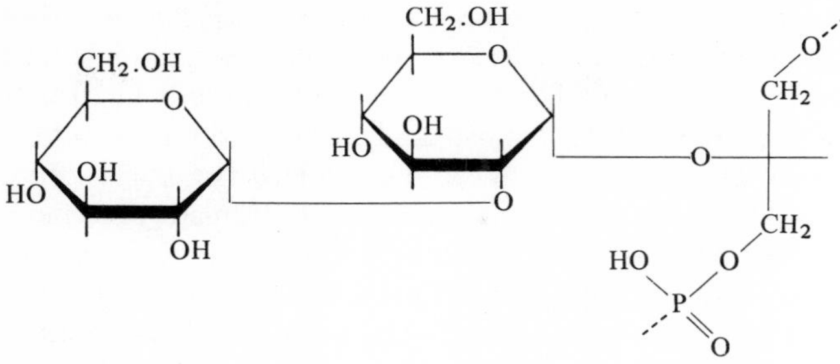

Fig. 5. Membrane teichoic acid from *Streptococcus faecalis* N.C.I.B. 8191.

sugars. In preparations from *Staphylococcus* species N.C.T.C. 7944 about one third of the glycerol residues are substituted by *N*-acetylgalactosaminyl residues at the 2-position.[32] Most of the glycosyl linkages have the α configuration, but there is evidence for the presence of a small number of β linkages. It is not known whether, like the α and β linkages in the teichoic acid from *Staphylococcus aureus*, these occur in separate molecules. In *Bacillus subtilis* N.C.T.C. 3610 all of the glycerol phosphate units are substituted by α-D-glucosyl residues.[33]

C. EXTENT AND REGULARITY OF SUBSTITUTION

The extent of glycosyl substitution on glycerol teichoic acids from various organisms shows a much wider variation than is the case for the ribitol teichoic acids where in most of the wild strains examined each ribitol possesses a glycosyl substituent. This raises the question whether such preparations are in fact mixtures of fully substituted and unsubstituted polymers or whether there is partial substitution on each polymer chain. The possibility also exists in the latter case whether in a partly substituted chain the substituents occur in a regular, irregular or even random manner. The first question has been answered differently in a number of cases.

Specific antisera have been prepared against teichoic acids where specificity depends upon the presence or absence of sugar residues on the chain. By using such antisera in precipitation experiments it has been possible to remove selectively from a mixture those components bearing glycosyl substituents. This procedure, together with biosynthetic arguments, has been used to show that preparations of wall teichoic acids containing ribitol or glycerol from several organisms are in fact mixtures of unsubstituted and fully substituted chains.[33-35]

In those cases that have been examined by chemical methods a different conclusion has been reached. Phosphodiesters are normally very stable towards alkali, but are labile if there is a hydroxyl group on the carbon atom adjacent to that bearing the phosphate. The lability of esters of this kind is due to their ability to form intermediate 5-membered cyclic phosphates with consequent fission of one of the P-O bonds of the diester. The cyclic phosphates are themselves labile towards alkali giving a mixture of isomeric phosphomonoesters. It is for this reason that RNA and polymers of glycerol phosphate or ribitol phosphate are easily hydrolysed to monoesters with alkali. A glycerol phosphate polymer with glycosyl substituents on each glycerol would not possess the necessary hydroxyl groups on carbon atoms adjacent to those involved in phosphodiester linkage and so would be unaffected by alkali even under vigorous conditions. Thus if a glycerol teichoic acid preparation, in which the molecular ratio of sugar: glycerol is less than one, consists of a mixture of unsubstituted and fully substituted polymers, then alkali would hydrolyse the unsubstituted polymer but would be without action on the fully substituted one. When the polymer from the *Staphylococcus* species N.C.T.C. 7944, which is partly substituted with *N*-acetylgalactosamine, is heated in alkali, hydrolysis is substantial, a major product being galactosaminylglycerol.[32] It follows that completely substituted polymer chains are absent, and the presence of partially substituted chains has been confirmed by the characterization of hydrolysis products in which substituted and unsubstituted glycerol residues were joined together through a phosphodiester linkage. It follows that not only are there no fully substituted chains, but there can be no significant number of situations where two *N*-acetylgalactosaminylglycerol phosphate units occur adjacent to each other. The unlikely possibility of a loss of a few sugar residues by acid hydrolysis during isolation has been eliminated in a similar type of teichoic acid in the walls of *Lactobacillus buchneri* N.C.I.B. 8007. In this teichoic acid about one third of the glycerol residues are substituted by α-D-glucopyranosyl.[36] The direct action of alkali on the cell wall gives a range of products closely similar to those from the staphylococcus but containing glucose rather than *N*-acetyl-galactosamine. As acid-catalysed loss of sugar could not occur in this case it is concluded[37] that a fairly uniformly partially glucosylated glycerol teichoic acid is present.

Examples of glycerol teichoic acids in which the sugar residues form a part of the main polymer chain are known. The representatives whose structures have been elucidated are simpler than are the ribitol teichoic acids that occur as pneumococcal capsular substances; they possess in their repeating structure only one or two simple sugar residues attached to glycerol phosphate. The first examples[38] were polymers of glucosylglycerol phosphate and galactosylglycerol phosphate occurring together in separate molecules in the wall of *Bacillus licheniformis* A.T.C.C. 9945; that containing glucose is illustrated in Fig. 6. In

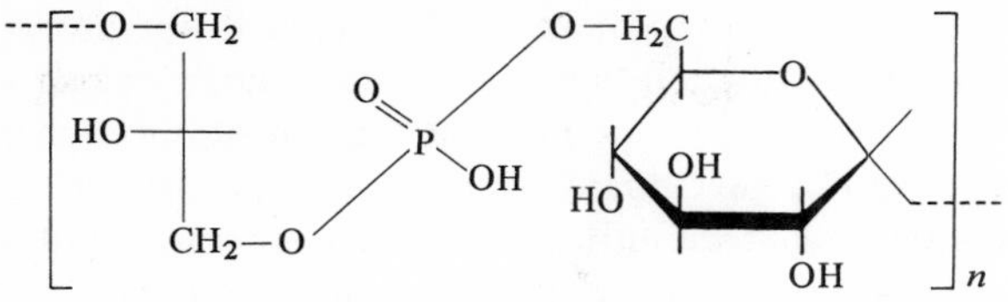

Fig. 6. Glucosylglycerol phosphate teichoic acid from *Bacillus licheniformis* A.T.C.C. 9945.

Lactobacillus plantarum N.I.R.D. C106 a mixture of related teichoic acids occurs;[39] in this case however the mixture is one of a polymer of glucosylglycerol phosphate and two polymers of isomeric diglucosylglycerol phosphates.[40] The characterization of this mixture has been studied by the use of

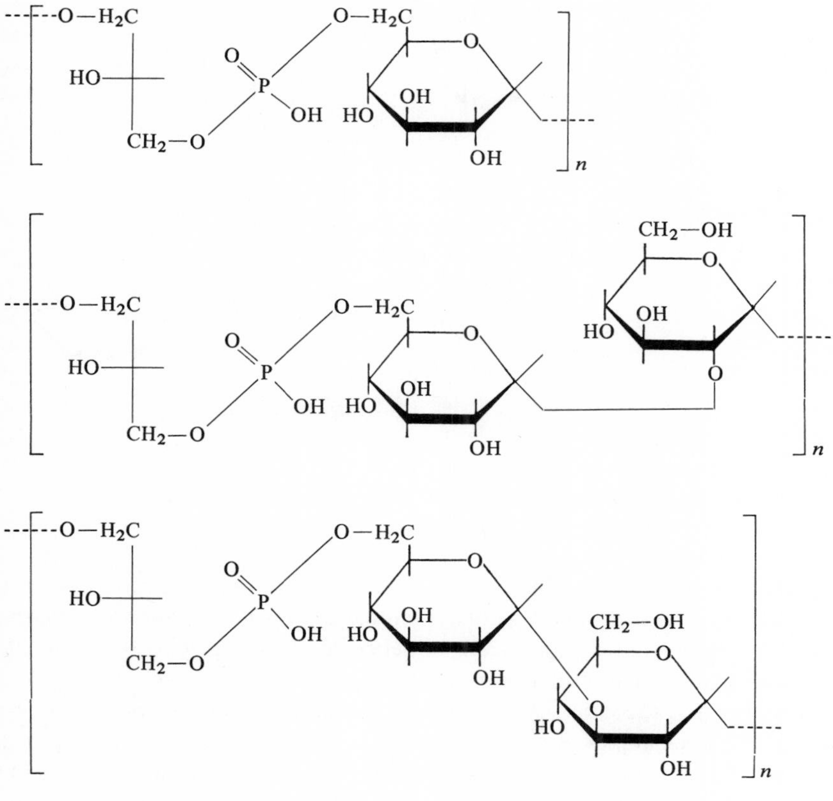

Fig. 7. Mixture of teichoic acids in walls of *Lactobacillus plantarum* N.I.R.D. C106.

the lectin concanavalin A, a protein from beans with the property of forming insoluble precipitates with polysaccharides and related compounds possessing α-D-glucosyl or α-D-mannosyl residues with unsubstituted hydroxyl groups at C-3, C-4 and C-6. The precipitate formed from the teichoic acid and lectin contains a polymer composed entirely of α-D-glucopyranosyl (1 → 2)-α-D-gluco-pyranosyl (1 → 1)-L-glycerol 3-phosphate; material containing repeating units of α-D-glucopyranosyl (1 → 3)-α-D-glucopyranosyl (1 → 1)-L-glycerol 3-phosphate remains in the supernatant and thus it is concluded that the different repeating units occur in at least two and probably three separate polymers. The probable composition of the teichoic acid mixture is given in Fig. 7.

Several glycerol teichoic acids have been described in which the glycerol phosphate residues are believed to be joined through 1,2 rather than 1,3-

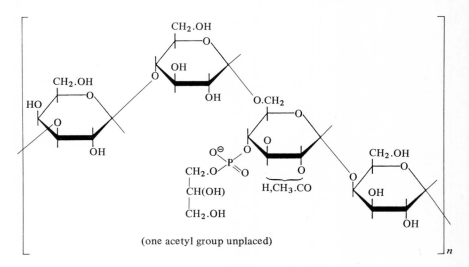

(one acetyl group unplaced)

Fig. 8. Polysaccharide from *Diplococcus pneumoniae* type 11A.

phosphodiester linkages. These have been found in actinomycetes (ref. 41, reviewed in ref. 16), and a similar example occurs in *Bacillus stearothermo-philus*.[42] However, the degradative evidence for this is inconclusive, and it is possible that the sugar residues form a part of the polymer chain.

Several pneumococcal capsular substances contain glycerol phosphate residues. In the examples so far examined in detail the molecules are poly-saccharides to which glycerol phosphate residues are attached as substituents to sugars through phosphodiester linkages.[43-46] In Fig. 8 the substance from *Diplococcus pneumoniae* type 11A is illustrated.[46]

D. SUGAR 1-PHOSPHATE DERIVATIVES

It has become apparent that the walls of a number of bacteria contain teichoic acids that do not conform to the pattern of structure shown in polymers of the polyol phosphates or glycosyl derivatives thereof; these are the representatives containing sugar 1-phosphate residues. Their structure is reflected in their high lability towards acid, and consequent hydrolytic decomposition during prolonged extraction from walls with acids; they also show characteristic instability under alkaline conditions, where hydrolysis of the 1-phosphate linkage exposes the reducing centres of the sugar residues to the action of alkali and consequent formation of saccharinic acids. The first example[47,48] was found in *Staphylococcus lactis* I3; it contains a repeating unit in which glycerol phosphate is attached to the hydroxyl at C-4 on *N*-acetylglucosamine 1-phosphate with D-alanyl on the hydroxyl at C-6. Its structure is shown in Fig. 9. A similar compound occurs in a related micrococcus[49] and there is evidence for the occurrence of related teichoic acids in other organisms.[50]

Other polymers containing sugar 1-phosphate residues occur in the walls of several bacteria. Although these lack polyol phosphate residues and so are strictly not teichoic acids, their occurrence, structure and properties are so similar that they should be included in discussions on teichoic acids. One that has received attention regarding its biosynthesis is a simple polymer of α-*N*-acetylglucosamine 1-phosphate in which the phosphate joins with the hydroxyl at C-6 on an adjacent unit.[50] It occurs in *Staphylococcus lactis* N.C.T.C. 2102. The walls of *Micrococcus* sp. A1 contain a polymer of 3-*O*-α-D-glucopyranosyl-*N*-acetylgalactosamine 1-phosphate in which linkage to the adjacent unit is through a phosphodiester at a hydroxyl on C-6 of glucose.[51] A polymer of similar or even possibly identical structure occurs, together with a conventional glycerol teichoic acid, in the walls of *Bacillus subtilis* 168 (M. Duckworth, A. R. Archibald & J. Baddiley, unpublished).

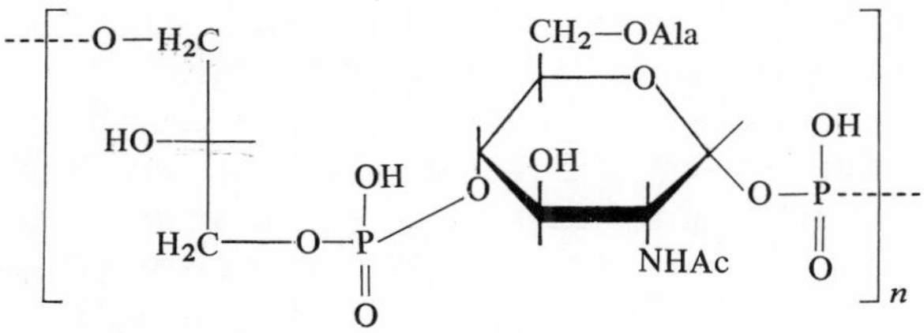

Fig. 9. Teichoic acid containing sugar 1-phosphate linkages from the walls of *Staphylococcus lactis* 13.

V. Linkage of Teichoic Acid to Wall and Membrane

A. LINKAGE WITH PEPTIDOGLYCAN

The slow rate of extraction of teichoic acids from walls by the use of trichloroacetic acid or alkali suggests that they are bound covalently to the peptidoglycan. Moreover, provided that the lytic enzymes that are frequently associated with walls are destroyed or prevented from hydrolysing the peptidoglycan, wall preparations can be washed exhaustively with water or buffers without the removal of the teichoic acid. Knowledge of the nature of this linkage is however far from complete. Preparations that have been extracted from walls with acid contain a small proportion of their phosphate as phosphomonoester, representing the ends of chains. If however the material is prepared by partial enzymic destruction of the peptidoglycan component of the wall, the resulting teichoic acid has attached to it small fragments of peptidoglycan and no phosphomonoester residues can be detected.[52,53] It is concluded that the teichoic acid is probably attached to the peptidoglycan through its terminal phosphate group in the form of a phosphodiester.

The attachment of an acidic polysaccharide to the wall of a lactobacillus is also believed to occur through a phosphodiester linking the peptidoglycan with the reducing end of the polysaccharide chain.[54,55] This conclusion is based on the rate of acid hydrolysis of the linkage and the simultaneous appearance of phosphomonoester groups. Moreover, acid hydrolysates of whole walls from many bacteria contain a phosphate of muramic acid,[56] and it is thought likely that this represents the linkage point between peptidoglycan and either teichoic acid or acidic polysaccharide.

A direct approach to a study of the structural details of the linkage between the macromolecular components of walls would require the partial destruction of both the teichoic acid chain and the glycan chain, and the preservation of the linkage together with recognizable components of both macromolecules still attached to this linkage component. This approach is difficult because it requires not only highly specific methods for the partial destruction of both polymer chains, but the necessarily small proportion of the wall material representing the linkage and its immediate environmental components must be clearly distinguished from contaminating material originating from small amounts of membrane, capsule, medium, etc. that are frequently very difficult to remove from walls. In a study along these lines using *Staphylococcus lactis* I3 the teichoic acid (Fig. 9) has been degraded by hydrolysis of acid-labile sugar 1-phosphate linkages, and the glycan has been degraded by autolytic enzymes adhering to the wall.[12] It is concluded that a glycerol residue is attached through a phosphodiester linkage to probably the hydroxyl on C-6 of a muramic acid in the peptidoglycan.

B. LINKAGE OF MEMBRANE TEICHOIC ACID WITH LIPIDS

When a suspension of bacterial cells is disrupted mechanically by any of the several methods available, the walls can be recovered by centrifugation at moderate speed, leaving fragmented membrane and cell contents in the supernatant. Centrifugation at high speed sediments the fragmented membrane together with the membrane teichoic acid.[15] This teichoic acid then appears to be associated with, or attached to, membrane or mesosomal residues. When the preparation procedure includes extraction from broken cells or supernatant with cold dilute trichloroacetic acid the teichoic acid is freely soluble in water and does not show the properties expected of a very large molecule. Thus it seems likely that membrane teichoic acid is covalently attached to a component of the membrane, and that acid extraction methods cause the hydrolysis of this linkage.

Membrane teichoic acids can be extracted from broken cells with a mixture of phenol and water, whereupon the teichoic acid is recovered from the aqueous phase.[57] The suggestion[2] that in such preparations the linkage to other membrane components should remain intact has been confirmed in a study on teichoic acids that have been extracted by this method from several lacto-bacilli.[58,59] Such preparations contain both teichoic acid and membrane lipids, and have been called "lipoteichoic acids". Although the lipid components could not be removed by organic solvents, and the preparations behaved as homogeneous entities in agar-gel diffusion studies using specific antisera, the

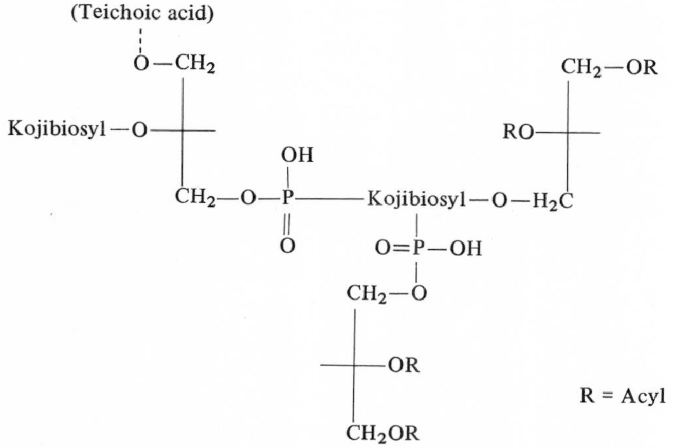

Fig. 10. Structure of the membrane lipoteichoic acid from *Streptococcus faecalis* N.C.I.B. 8191.

nature of the association between lipids and teichoic acid was not established. In studies on the lipoteichoic acid from *Streptococcus faecalis* N.C.I.B. 8191 it has been shown[31] that all of the membrane teichoic acid occurs in this form. It was found that the glycerol phosphate chain of the teichoic acid comprises 28-35 repeating units and is covalently attached through its terminal phosphate to a phosphatidyl glycolipid that is a normal component of the membrane. The phosphodiester linkage is to an unidentified hydroxyl on one of the two glucose residues of the glycolipid, and the phosphatidyl residue is attached to another unidentified hydroxyl group on a glucose; the structure is illustrated in Fig. 10. It seems likely that membrane teichoic acids in other bacteria are attached to glycolipids in their membranes.

VI. Biosynthesis

The present day knowledge of the role of nucleotide precursors in the biosynthesis of glycosides, polysaccharides and related natural products might suggest immediately that such precursors would participate in the synthesis by bacteria of teichoic acids and sugar 1-phosphate polymers. This would seem

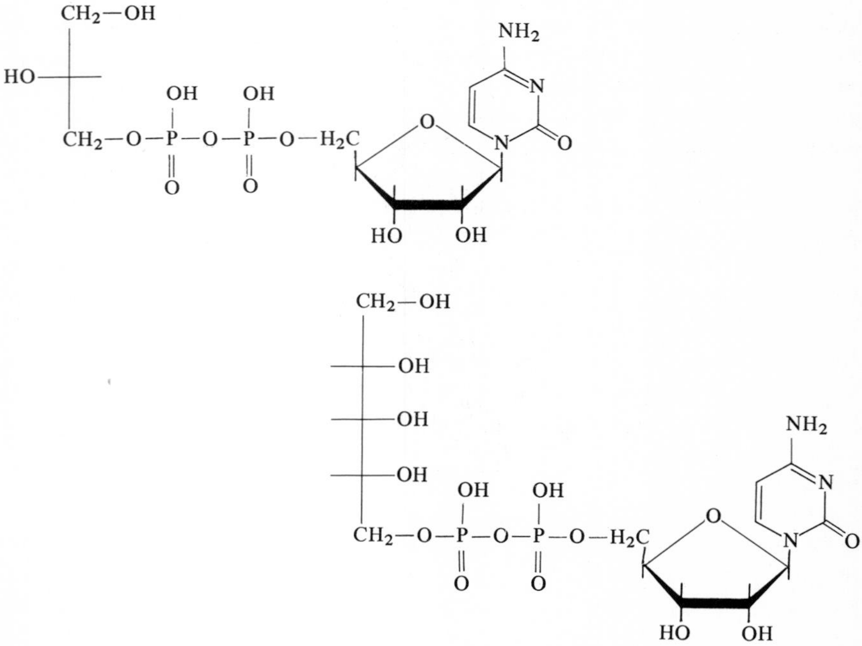

Fig. 11. Cytidine diphosphate glycerol and cytidine diphosphate ribitol, precursors of the teichoic acids.

likely not only for the introduction of glycosyl substituents into these compounds, but also for the building of chains of polyol phosphate units. In fact, the reverse of this kind of reasoning occurred at the time of the discovery of the teichoic acids. The nucleotides cytidine diphosphate glycerol and cytidine diphosphate ribitol (see Fig. 11) had been isolated from bacterial extracts,[60,61] their structures determined[62-64] and synthesized[65,66] before the discovery of the teichoic acids. Enough was known at the time to predict that these nucleotides, although differing from the nucleoside diphosphate sugars referred to in the analogy, might nevertheless represent precursors of larger molecules containing either the polyols ribitol and glycerol or their phosphates. Thus it was predicted that bacteria should contain polymers of glycerol phosphate or ribitol phosphate and that CDP-glycerol and CDP-ribitol would be their biosynthetic precursors. This prediction resulted in the discovery[1] of the teichoic acids, and these polymers are probably unique in that their biosynthetic precursors were known and had been studied in considerable detail before the products themselves had been discovered.

A. NUCLEOTIDE PRECURSORS

The suggested role of the nucleotides CDP-glycerol and CDP-ribitol in the transfer of polyol phosphate residues to the growing chains of teichoic acids is in agreement with the established stereochemistry of the residues in the nucleotides and with the corresponding stereochemistry of these residues in the polymer chains. For example, in those cases where the configuration of the ribitol phosphate residues in a conventional teichoic acid or pneumococcal capsular substance has been established it is found that they possess the D-5 configuration, identical with that of the ribitol phosphate in CDP-ribitol.

Gram-positive bacteria have been shown[67] to contain the enzymes necessary to synthesize CDP-glycerol from D-glycerol 1-phosphate and CTP (CDP-glycerol pyrophosphorylase), and CDP-ribitol from D-ribitol 5-phosphate and CTP (CDP-ribitol pyrophosphorylase). It is assumed that glycolysis is the source of the glycerol phosphate, which is structurally identical to that participating in the biosynthesis of phospholipids. The ribitol phosphate arises through enzymic reduction by NADH of D-erythropentulose 5-phosphate (D-ribulose 5-phosphate).[68]

Washed cytoplasmic membranes, or particulate preparations corresponding to fragmented membrane, are able to catalyse the synthesis of poly(glycerol phosphate) from CDP-glycerol.[57] The first successful experiments were carried out with preparations from bacilli, but similar preparations from other bacteria have been used in more recent studies. A variety of methods has been used for making the enzyme preparations including sonic oscillation, French pressure cells, mechanical vibrators, grinding with alumina and enzymic dissolution of cell walls. As these preparations often contain teichoic acids, it has not been possible to determine whether a primer, possibly containing the linkage with peptidoglycan, is necessary or whether new chains are formed. A scheme for the route to the synthesis of poly(glycerol phosphate) is given in Fig. 12.

D-Glycerol 1-phosphate + CTP $\rightleftharpoons$ CDP-glycerol + PP_i

CDP-glycerol + (Glycerol phosphate)$_n$ $\rightarrow$ (Glycerol phosphate)$_{n+1}$ + CMP

Fig. 12. Synthesis of CDP-glycerol and poly(glycerol phosphate)

The particulate enzyme system from *Bacillus subtilis* N.C.T.C. 3610 which effects synthesis of poly(glycerol phosphate) is also able to transfer glucose residues from UDP-glucose to the free hydroxyl groups of the polymer,[33] giving a teichoic acid in which all glycerol phosphate residues are substituted by α-D-glucopyranosyl. The substrate for this transglycosylation can be either poly(glycerol phosphate) that has been extracted from the cell wall of the organism, or polymer that has been synthesized *in vitro*. Like the polymerization reaction, the transglycosylation requires for its optimum rate a high concentration of either Ca^{2+} or Mg^{2+}. Efforts to solubilize the two enzymes concerned in the synthesis of polymer from nucleotide precursors have been unsuccessful, and it is concluded that these and other enzymes concerned in teichoic acid syntheses are integral parts of the membrane. On the other hand, CDP-glycerol pyrophosphorylase and CDP-ribitol pyrophosphorylase, like the nucleoside diphosphate sugar pyrophosphorylases, are water-soluble enzymes. Nevertheless, it can be shown with carefully prepared membrane particles that these soluble enzymes are in fact mainly loosely associated with the membrane (unpublished observations cited in ref. 2).

Essentially similar enzyme preparations representing fragmented membrane catalyse the synthesis of poly(ribitol phosphate) from CDP-ribitol. The synthesis has been studied in detail with *Lactobacillus plantarum* A.T.C.C. 8014 (ref. 69) and with *Staphylococcus aureus* Copenhagen.[70] In both cases the properties of the enzyme system are similar to those of the system that synthesizes poly(glycerol phosphate); CMP is produced and high concentrations (about 20 mM) of bivalent cations are required for optimal synthesis. Particulate enzyme preparations from the latter organism catalyse the transfer from UDP-*N*-acetylglucosamine of glycosyl residues to pre-formed poly(ribitol phosphate), but the efficiency of transfer is considerably greater if CDP-ribitol rather than polymer is used. Nevertheless, β-*N*-acetylglucosaminyl residues can be removed from a teichoic acid preparation that possesses both α and β linkages by the use of a β-*N*-acetylglucosaminidase, and it is then found that both α and β-*N*-acetylglucosaminyl residues can be re-inserted by the enzyme preparation and UDP-*N*-acetylglucosamine in a ratio approximately corresponding to that in the original mixture.[71,72] Glucosyl transfer from UDP-glucose to poly(ribitol phosphate) with enzyme preparations from *Bacillus subtilis* W-23 follows a similar course, but in this case the linkages are entirely of the β-configuration.[35]

Obviously the sequence of events in the synthesis of teichoic acids in which the sugar residues form a part of the main polymer chain (e.g. Figs 6, 7, 9) must differ from those already discussed where the main chain is a simple polyol phosphate polymer. Although it has been shown with the glucosylglycerol phosphate (Fig. 6) and galactosylglycerol phosphate teichoic acids from *Bacillus licheniformis* that both CDP-glycerol and respectively UDP-glucose or UDP-galactose are required together for synthesis of polymer,[38] the details of the biosynthetic process, including especially the participation of lipid intermediates, were not understood at the time and so the interpretation of some of the observations is difficult. For the synthesis of the wall teichoic acid from *Staphylococcus lactis* I3 (Fig. 9) using a cell-free system[73] both CDP-glycerol and UDP-*N*-acetylglucosamine are required simultaneously, and these nucleotides provide respectively a glycerol phosphate and an *N*-acetylglucosamine 1-phosphate residue (see Fig. 13). The transfer from a nucleotide of a sugar

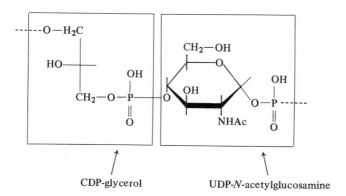

CDP-glycerol   UDP-*N*-acetylglucosamine

Fig. 13. Transfer of residues from nucleotide precursors to teichoic acid from *Staphylococcus lactis* I3.

1-phosphate residue is presumably a general mechanism for the synthesis of polymers containing this feature in their structure, and has been shown[74] to occur in the biosynthesis of poly(*N*-acetylglucosamine 1-phosphate) from UDP-*N*-acetylglucosamine using fragmented membrane preparations from *Staphylococcus lactis* N.C.T.C. 2102.

B. ALANINE ESTER RESIDUES

It has been known for some time that Gram-positive bacteria possess a soluble enzyme with high specificity for D-alanine which catalyses a reaction between the amino acid and ATP; an alanyl-AMP-enzyme complex has been assumed to be formed together with inorganic pyrophosphate.[75] The complex reacts spontaneously with hydroxylamine, and the enzyme differs from that which participates in the introduction of D-alanine into peptidoglycan. Nevertheless, although it was thought likely that it is concerned in the introduction of alanyl ester residues into teichoic acids, efforts to use it or other preparations for the attachment of alanyl ester residues to either polymers or precursors have failed. More recently the problem has been re-examined.[76] It is found that the incorporation of D-alanine into isolated membrane of *Lactobacillus casei* A.T.C.C. 7469 requires ATP, Mg^{2+} and a soluble enzyme that appears to differ in some respects from the alanine activating enzyme. The soluble enzyme has been called D-alanine:membrane ligase and the acceptor is probably the membrane teichoic acid, although isolated teichoic acid is not affected by the system. It is possible that the enzyme requires precise orientation of the acceptor membrane teichoic acid within the membrane, and the glycolipid component of this teichoic acid may also be required. The relationship between the ligase and the D-alanine activating enzyme is not yet clear.

C. LIPID INTERMEDIATES

In order to understand the details of the biosynthetic route to the teichoic acids, in particular the role of lipid intermediates, it is necessary to consider the mechanism of synthesis of other components of the outer layers of bacteria, especially the synthesis of peptidoglycan. It is well-established that the synthesis of glycan chains is a transglycosylation process in which the precursors are UDP-*N*-acetylglucosamine and UDP-*N*-acetylmuramyl peptide, where the peptide bears a structural relationship to the corresponding peptide chains in the final polymer. The synthesis of peptidoglycan occurs for the most part in the cytoplasmic membrane, and cell-free particulate enzyme preparations have been used to demonstrate[77] that residues are transferred from the nucleotide

$$H-\left(CH_2-\overset{\overset{\displaystyle CH_3}{|}}{C}=CH-CH_2\right)_{11}-O-\overset{\overset{\displaystyle OH}{|}}{\underset{\underset{\displaystyle O}{\|}}{P}}-OH$$

Fig. 14. Undecaprenol phosphate

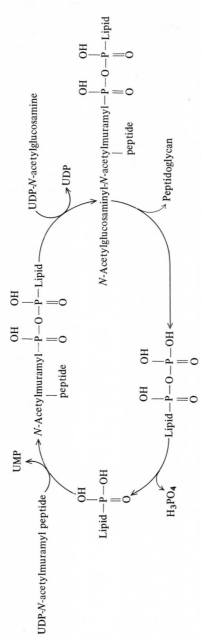

Fig. 15. Biosynthetic pathway to peptidoglycan through lipid intermediates

precursors to a lipid phosphate, identified in several cases as undecaprenol phosphate (Fig. 14). The first step in the route, illustrated in Fig. 15, is the transfer to the undecaprenol phosphate of an N-acetylmuramyl peptide monophosphate residue, giving the corresponding substituted lipid pyrophosphate and UMP. An N-acetylglucosamine residue is then transferred directly to this product with the formation of UDP. Thus the lipid now possesses a disaccharide unit corresponding to the repeating disaccharide structure in peptidoglycan. At this point modifications may be made to the peptide chain, including the attachment of "bridging" peptides, e.g. the pentaglycine bridge in *Staphylococcus aureus*. The disaccharide-peptide is then transferred to an acceptor, which is presumably a growing end of a glycan chain. The other product of the transfer is undecaprenol pyrophosphate, which is monodephosphorylated by a specific phosphatase thereby completing the lipid phosphate cycle. Peptidoglycan synthesis is completed outside the membrane by a cross-linking transpeptidation in which a linkage is formed between a penultimate D-alanyl carboxyl group of one peptide chain and, in the case of *S. aureus*, a glycine amino group on the bridging peptide of a neighbouring chain; the terminal D-alanyl residue of the original chain is thereby expelled as alanine.

Similar lipid phosphate intermediates occur in the synthesis of teichoic acids and related polymers. One of the simplest demonstrations of this is given by the synthesis of the poly(N-acetylglucosamine 1-phosphate) that occurs in the walls of *Staphylococcus lactis* N.C.T.C. 2102. In this polymer the phosphodiester linkage is between hydroxyl groups at positions 1 and 6 in adjacent units. The only nucleotide required is UDP-N-acetylglucosamine, and by using this with an isotopic label in either the sugar, the phosphate or both, it can be shown that the sugar 1-phosphate residue is transferred from the nucleotide to an intermediate that is soluble in butan-1-ol or other organic solvents. Pulse-labelling techniques demonstrate the transfer from nucleotide to lipid intermediate and finally to polymer.[78] The intermediate is a derivative of α-N-acetylglucosamine 1-pyrophosphate and, although the exact nature of the lipid has not been established by direct methods, it is believed to be an isoprenoid. The biosynthetic route for this polymer is illustrated in Fig. 16, and although there is a considerable similarity between this and the route to the biosynthesis of peptidoglycan, it should be noted that the presence of phosphate in the final product ensures that lipid monophosphate is returned directly to the cycle after transfer of the N-acetylglucosamine 1-phosphate residue to the growing polymer chain.

The participation of lipid intermediates in the biosynthesis of other wall polymers has been established.[79] Thus they have been detected in the synthesis by particulate enzyme preparations of the wall teichoic acid of *Staphylococcus lactis* I3 (Fig. 9). UDP-N-acetylglucosamine donates an N-acetylglucosamine 1-phosphate residue to a lipid phosphate to give an intermediate that can accept a glycerol phosphate residue from CDP-glycerol. The product now possesses a

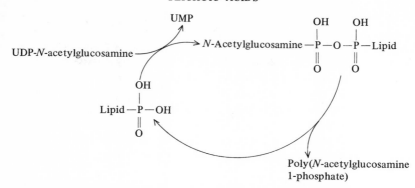

Fig. 16. Lipid intermediates in the synthesis of poly(N-acetylglucosamine 1-phosphate).

complete repeating unit, and pulse-labelling experiments support the route for the synthesis outlined in Fig. 17. In agreement with this scheme, the addition of CDP-glycerol to yield the intermediate containing the complete repeating unit is accompanied by a corresponding decrease in the amount of the first intermediate.[80]

An interesting variation is provided by *Bacillus licheniformis* A.T.C.C. 9945. The wall teichoic acids in this organism include the polymer of glycerol

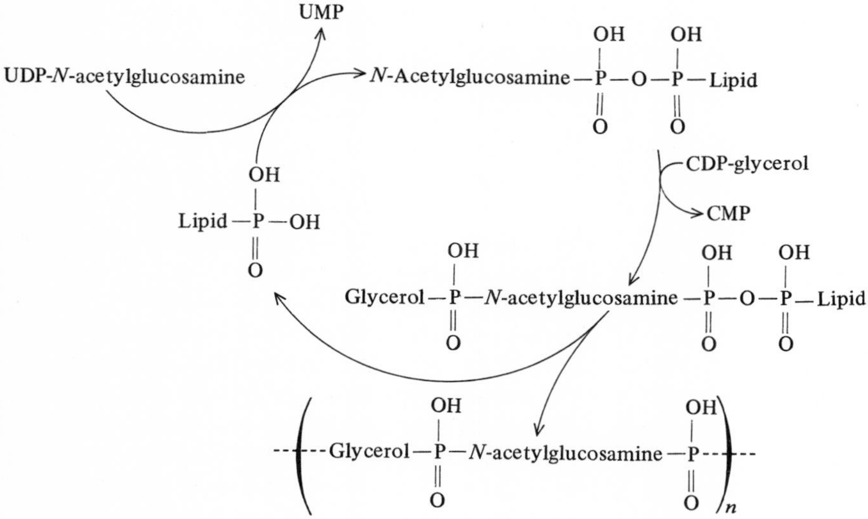

Fig. 17. Lipid intermediates in the synthesis of the wall teichoic acid from *Staphylococcus lactis* 13.

phosphate and glucose (Fig. 6) in which the glucose forms a part of the main polymer chain. In the synthesis of this polymer by a cell-free preparation it can be shown[81] that the first-formed intermediate from UDP-glucose is a lipid containing glucose. None of the phosphate in this lipid is derived from the nucleotide, and it is concluded that it is a glucosyl-phosphate-lipid with only one phosphate group. When CDP-glycerol is present, the amount of this intermediate decreases and a new intermediate results in which glycerol phosphate and glucose are present in the structure now forming the repeating unit of the polymer. Pulse-labelling experiments support the order of formation of the two intermediates and the biosynthetic route is given in Fig. 18. This synthesis

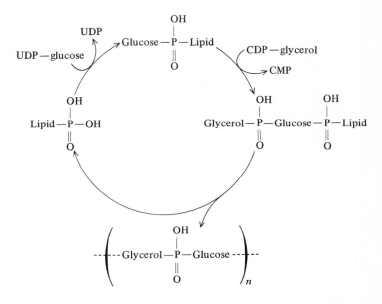

Fig. 18. Transglycosylation through lipid intermediates in the synthesis of a wall teichoic acid in *Bacillus licheniformis*

differs from the others in that the intermediates do not contain a pyrophosphate structure, and the overall reaction is a transglycosylation rather than a transphosphorylation, i.e. the linkage formed on the transfer of a repeating structure to the growing polymer is a glycosidic linkage between glucose and a hydroxyl on glycerol. Another difference between this synthesis and those involving transfer of phosphate is that, whereas in the nucleotide precursor UDP-glucose the glucosyl linkage is α, in the lipid intermediate it has the β-configuration. In the teichoic acid itself the glucosyl residues have the α-configuration, and so the overall transfer from nucleotide to polymer constitutes a double inversion. A similar double inversion has been observed in the introduction of glucose residues into lipopolysaccharide from UDP-glucose where a glucose-phosphate-lipid participates.[82]

The synthesis of simple polymers of glycerol phosphate, as found in the wall of *Bacillus licheniformis*, also proceeds through a lipid intermediate.[83] Although the structure of the intermediate has not been studied in detail, it is believed to be a glycerol-pyrophosphate-lipid which transfers glycerol phosphate to the growing polymer chain.

D. NATURE OF THE LIPID

The lipid phosphate that participates in the synthesis of peptidoglycan has been shown in a few cases to be undecaprenol phosphate,[77,84] and the same lipid phosphate participates in the synthesis of the oligosaccharide chains of lipopolysaccharide[85] and of a bacterial mannan.[86] Identification in all cases has been achieved by isolation and purification of the phosphate followed by a study of chemical properties and mass spectroscopy. Although the lipid that participates in the synthesis of teichoic acids and related polymers also exhibits some of the chemical properties expected of a polyisoprenoid, technical difficulties have so far prevented the isolation of the large amounts of cytoplasmic membrane that would be required for a similar characterization of this compound. Nevertheless, the identity of this lipid has been established in two cases by an indirect method.

The study is based on the conclusion that if the lipid phosphate that participates in the synthesis of peptidoglycan is the same as that for the synthesis of teichoic acid, and if there is a common pool of this phosphate for both syntheses, then by using a cell-free system capable of synthesizing both polymers it should be possible to demonstrate an interdependence of the two syntheses. The synthesis of peptidoglycan and wall teichoic acid (Fig. 6) in particulate membrane preparations from *Bacillus licheniformis* A.T.C.C. 9945 has been studied in this connection,[83] where it is found that the rate of synthesis of peptidoglycan is markedly decreased when nucleotide precursors for teichoic acid are present. Conversely, the rate of synthesis of teichoic acid is decreased in the presence of nucleotide precursors for peptidoglycan. The optimum conditions of temperature and pH are not identical for the two syntheses, and maximum decreases in rate were achieved by adjusting the conditions to be at the optimum for the competing synthesis.

Further confirmation of the interdependence of the two systems can be obtained by studying the effects of the antibiotic bacitracin. This antibiotic powerfully inhibits bacterial growth by interfering with the synthesis of peptidoglycan. In fact, it specifically inhibits the dephosphorylation of undecaprenol pyrophosphate to the monophosphate, thereby blocking the

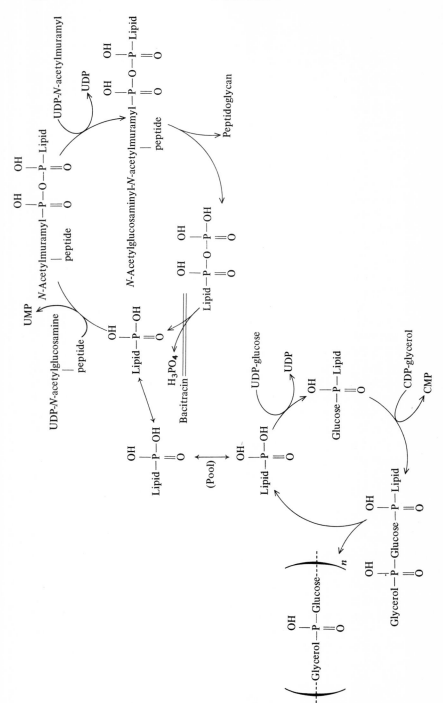

Fig. 19. Simultaneous synthesis of teichoic acid and peptidoglycan in cell-free system from *Bacillus licheniformis*

completion of the lipid phosphate cycle.[87] In the biosynthesis of teichoic acids lipid pyrophosphate is not an intermediate, and so bacitracin has no inhibitory action in a cell-free system. On the other hand, when both polymers are synthesized simultaneously, the antibiotic would prevent the return of lipid monophosphate to the common pool and consequently cause the accumulation of lipid intermediates in the peptidoglycan cycle. This would deprive the teichoic acid cycle of lipid phosphate and inhibition of teichoic acid synthesis should occur. Such an inhibition is actually observed, and it is concluded that both syntheses make use of the same undecaprenol phosphate molecules.[83] The interdependent biosynthetic routes sharing lipid phosphate are given in Fig. 19. In the study described, it has been shown that the synthesis of a second wall teichoic acid, poly(glycerol phosphate), exhibits interdependence with the synthesis of the other two polymers and it follows that all these syntheses share undecaprenol phosphate. In a related study with a staphylococcus[88] the same conclusion has been reached.

E. DIRECTION OF CHAIN-EXTENSION

There are two known methods of chain-extension in the synthesis of natural macromolecules. Glycan chains in glycogen are extended by apparently direct transfer of glucose units from nucleoside diphosphate sugar to the non-reducing end of chains, whereas the synthesis of peptide chains in proteins occurs at the carboxyl end by transfer of the growing oligopeptide from its compound with RNA to an aminoacyl-RNA. The synthesis of oligosaccharide chains in lipopolysaccharide occurs through undecaprenol phosphate intermediates, and it is found[89] that chain-extension occurs in a manner analogous to that in protein synthesis, i.e. by extension from the "reducing end" of the oligosaccharide. Thus the chain, which is built up from trisaccharide units, is transferred from an oligosaccharide-pyrophosphate-lipid to the terminal sugar residue of a trisaccharide-pyrophosphate-lipid (Fig. 20).

In order to determine the direction of growth of a chain pulse-labelling techniques are of great value. In the case of the oligosaccharide chains of lipopolysaccharide it can be shown that the last trisaccharide unit to enter the main chain is that possessing a reducing galactose residue in the product, and so growth must have occurred at that reducing end. The synthesis of poly(glycerol phosphate) chains in a teichoic acid has been examined by a similar method.[90] In a poly(glycerol phosphate) one end of the molecule should bear a phosphomonoester residue or a linkage with the peptidoglycan, whereas at the other end there should be a glycol structure. By the use of pulse-labelling of glycerol residues during synthesis, followed by oxidation of the glycol structure with periodate to give formaldehyde, it can be shown that the most recently introduced glycerol phosphate unit is at the "glycol end" of the chain (Fig. 21).

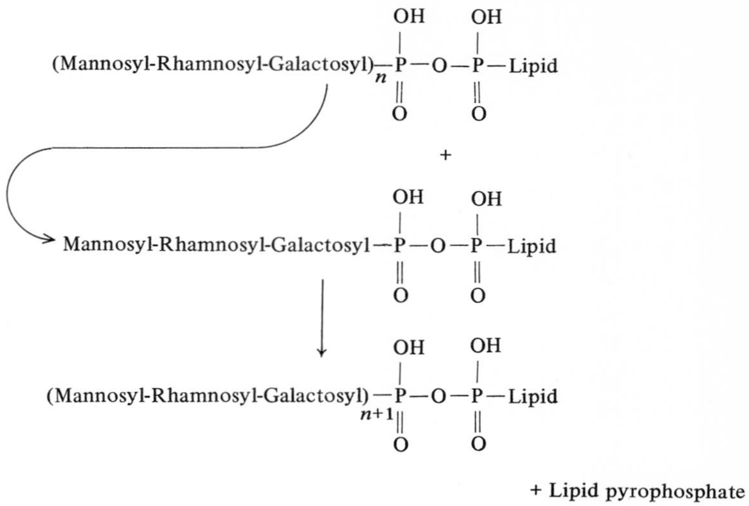

$$(\text{Mannosyl-Rhamnosyl-Galactosyl})_{n}\!\!-\!\!\overset{\overset{\displaystyle OH}{\displaystyle |}}{\underset{\underset{\displaystyle O}{\displaystyle \|}}{P}}\!\!-\!\!O\!\!-\!\!\overset{\overset{\displaystyle OH}{\displaystyle |}}{\underset{\underset{\displaystyle O}{\displaystyle \|}}{P}}\!\!-\!\!\text{Lipid}$$

+

$$\text{Mannosyl-Rhamnosyl-Galactosyl}\!-\!\overset{\overset{\displaystyle OH}{\displaystyle |}}{\underset{\underset{\displaystyle O}{\displaystyle \|}}{P}}\!\!-\!\!O\!\!-\!\!\overset{\overset{\displaystyle OH}{\displaystyle |}}{\underset{\underset{\displaystyle O}{\displaystyle \|}}{P}}\!\!-\!\!\text{Lipid}$$

$$(\text{Mannosyl-Rhamnosyl-Galactosyl})_{n+1}\!-\!\overset{\overset{\displaystyle OH}{\displaystyle |}}{\underset{\underset{\displaystyle O}{\displaystyle \|}}{P}}\!\!-\!\!O\!\!-\!\!\overset{\overset{\displaystyle OH}{\displaystyle |}}{\underset{\underset{\displaystyle O}{\displaystyle \|}}{P}}\!\!-\!\!\text{Lipid}$$

+ Lipid pyrophosphate

Fig. 20. Chain-extension of oligosaccharide in lipopolysaccharide

Thus chain-extension is in the opposite sense from that in the lipopoly-saccharide.

Similar methods have been used to study this problem with the poly(N-acetylglucosamine 1-phosphate) and the glycerol teichoic acid (Fig. 9) in staphylococci.[74,91] In the former case gentle acid hydrolysis of the sugar 1-phosphate linkages in the polymer gives N-acetylglucosamine from one end of the polymer and its 6-phosphate from the residues within the chain. It is found in these cases also that chain-extension occurs at the end of the chain which does not bear the phosphate.

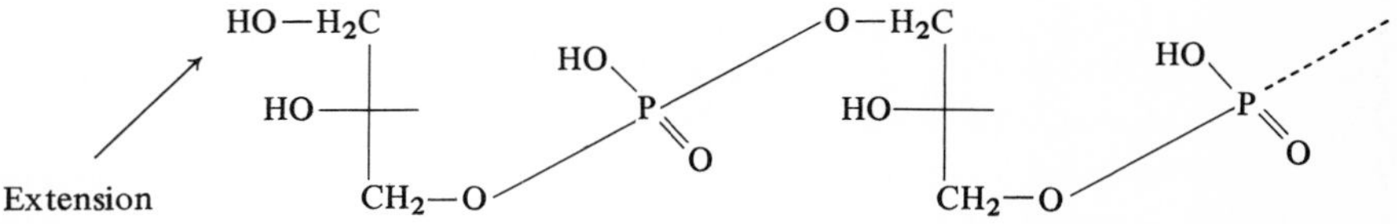

Fig. 21. Growth of a poly(glycerol phosphate)

F. CONTROL OF WALL SYNTHESIS

The regular and orderly growth and cell division of bacteria require the operation of accurate control processes in the regulation of the amount and composition of wall material. Experiments on living organisms which could help to clarify this problem are lacking but recent work with cell-free systems suggests how such control might be achieved. In the study[83] on the interrelationship between the synthesis of teichoic acids and peptidoglycan in a fragmented membrane system (see section D and Fig. 19) the decrease in rate of synthesis of one polymer which can be brought about by simultaneous synthesis of the other indicates that the availability of a common component must be rate-limiting for the two systems. As the only common component is undecaprenol phosphate it is concluded that the amount of this lipid phosphate controls the rate of synthesis of wall polymers in the cell-free system.

It is possible that in a living organism the factors effecting control of rate of wall synthesis might be different from those in the cell-free system, but it is reasonable to believe that rate control by regulating the amount of a single membrane component (lipid phosphate) would be a most attractive method for a cell to adopt. This suggestion[83] is supported by the observation that bacteria possess two enzymes[92] that are able to control the amount of undecaprenol phosphate in the membrane, a kinase for the phosphorylation of undecaprenol, and a phosphatase for its hydrolysis. Thus the amount of phosphate could be controlled by regulating the amounts or activities of either or both of these enzymes.

It is noteworthy that control of wall synthesis by regulating the amount of lipid phosphate would not itself effectively determine the composition of the wall. The observations that wall composition, i.e. the relative amounts of teichoic acid or acidic polysaccharide, may depend upon the composition of the growth medium and other external factors (see VII A & B), and that changes in wall composition can be brought about rapidly, indicates that sensitive control mechanisms must exist for such regulation. The mechanism of this control of composition is not yet understood, but it is possible that allosteric effects with the enzymes required for the synthesis of nucleotide precursors might be concerned. In this connection it is interesting that in a cell-free system the synthesis of CDP-glycerol from CTP and glycerol phosphate with the soluble synthetase is powerfully inhibited by UDP-*N*-acetylmuramyl peptide, a precursor for the synthesis of peptidoglycan (R. J. Anderson, H. Hussey & J. Baddiley, unpublished work).

If, in the sharing of undecaprenol phosphate for synthesis of teichoic acids and peptidoglycan, the lipid phosphate entered the common pool after each turn of the cycles, then the effect of bacitracin should be rapid and complete during a single cycle. In fact, the effect is less than would be expected. This suggests that there is some form of restriction on the return of lipid phosphate to the common pool. Moreover, in the synthesis of teichoic acid, pre-incubation of the system with nucleotide precursors for teichoic acid synthesis decreases the inhibiting action shown by addition of nucleotides for peptidoglycan synthesis. Similarly, in peptidoglycan synthesis, pre-incubation with nucleotide precursors for the synthesis of peptidoglycan decreases the inhibitory action of nucleotides for teichoic acid synthesis. These and related experiments support the view that return of lipid phosphate to the common pool may not occur after each turn of the cycle of introduction of repeating units to either the teichoic acid or the peptidoglycan chains. A possible interpretation of these observations[83] is that lipid phosphate remains associated with the enzyme system for the synthesis of a polymer until the whole chain has been completed, and only then is it returned to the common pool. This implies that the lipid phosphate and its derivatives do not move around freely in the highly organized and orientated molecular environment of the membrane; moreover, it is suggested that the multi-enzyme systems responsible for the synthesis of the several wall polymers are in fairly close proximity to each other for the purpose of sharing of lipid phosphate. This high degree of order in the separate systems in the membrane is consistent with the recent observation (see p. 39) that, at least in some organisms, a large proportion of the wall material comprises peptidoglycan and teichoic acid chains joined together in a 1:1 molecular ratio. Clearly, the synthesis of such a uniform molecular species requires a considerable degree of order and interdependence of the enzyme systems.

VII. Function

It is reasonable to assume that the possession of wall and membrane teichoic acids confers certain advantages upon bacteria, i.e. they probably have one or more functions. As both walls and membranes have a number of important purposes, it is likely that these major components participate in some way in the well-being of the cell; this seems especially likely in view of the marked physical and physiological properties of teichoic acids in walls and membranes discussed below. It is important, however, to distinguish between true biological functions of wall polymers and those of their properties that cause walls to behave in certain ways. For example, teichoic acids and wall polysaccharides frequently possess characteristic serological properties that are responsible for the agglutination of bacterial suspensions in the presence of specific antisera (see p. 38). Important though this property of teichoic acids is, it is on the whole unlikely that bacteria derive benefit from such behaviour, and in fact agglutination is normally regarded as a phenomenon with considerable disadvantage to micro-organisms, as this is in most cases the first step towards the phagocytosis of the organism by the host.

The sites of attachment of bacteriophages to the surface of Gram-positive bacteria are intimately related to the presence of teichoic acids or other polysaccharides in the outer region of the cell. The receptor sites appear to be organized structures involving both peptidoglycan and teichoic acids,[93-99] and the importance of the teichoic acid component has been established. Here again however it is unlikely that this involvement can be regarded as a useful function of teichoic acids, because infection of a cell by bacteriophage is mainly a disadvantageous phenomenon to the cell. Nevertheless, bacteria can sometimes derive advantage from such infection, as this can be a means of interchange of genetic material.

On balance, however, it is unlikely that either the serological properties of teichoic acids or their role in bacteriophage-receptor sites are primary functions of these molecules.

A. IMPORTANCE OF TEICHOIC ACIDS

The importance of teichoic acids to bacteria is indicated by their widespread distribution in the bacterial kingdom. Wall teichoic acids, including the structurally related polymers containing sugar 1-phosphate residues, are common among the Gram-positive bacteria but are nevertheless absent from a number of species. In many of those lacking wall teichoic acids, acidic polysaccharides possessing uronic acid residues (teichuronic acids) are present in the walls, and it is likely that these serve the same purpose. On the other hand, extensive studies with many Gram-positive bacteria indicate that membrane teichoic acids are always present in appreciable amounts. Although these polymers are not found in typical Gram-negative bacteria it is probably significant that these organisms possess a closely related class of macromolecule, the lipopolysaccharides, and it is likely that they serve a similar purpose. Lipopolysaccharides include in their structure phosphodiester groups, sugar residues and basic centres, and they are situated in the outer membrane of the cell.

The importance to microorganisms of teichoic acids can be demonstrated experimentally by studying the behaviour of bacilli, especially when grown in continuous culture under different carefully defined conditions. It has been shown[100] with *Bacillus subtilis* that, when grown in a chemostat with phosphate limitation of growth, teichoic acid is no longer present in the walls, but in its place is found a polysaccharide containing uronic acid residues. The substitution of the polymers occurs rapidly and is not only confined to newly formed cells. This shows that the organism is able to dispense with its wall teichoic acid, provided that it can replace it by another acidic polymer. It is especially significant that even under the conditions of phosphate limitation of growth a membrane teichoic acid is still produced,[101] suggesting a vital and indispensable role for this cell component. Thus it seems that cells must possess membrane teichoic acids, but all that is required in the wall is one or more acidic polymers.

If teichoic acids really are indispensable then it would be reasonable to expect that the inhibitory action of certain antibiotics might be ascribed to interference with teichoic acid synthesis. It has already been described how bacitracin can inhibit the synthesis of teichoic acid in a cell-free system when peptidoglycan synthesis is occurring simultaneously (see p. 61), but this is not a direct inhibitory effect. On the other hand, novobiocin inhibits the synthesis of teichoic acids in several cell-free systems.[57,69,70] In this case inhibition occurs independently of other syntheses and at concentration ratios of antibiotic to membrane comparable to the ratio of concentration of novobiocin to cells required to inhibit cell growth.[102] Thus it is possible that this antibiotic acts through inhibition of synthesis of teichoic acid, but it may also inhibit other cellular processes. Chloramphenicol at concentrations somewhat higher than those required to inhibit protein synthesis inhibits the incorporation of glucose into the teichoic acid in the walls of *Bacillus licheniformis*,[103] but it is unlikely that this interesting effect is of importance in our understanding of either the antibiotic action of chloramphenicol or the function of teichoic acids.

B. CATION BINDING

With the exception of the true halophiles, bacteria that are able to tolerate moderate concentrations of sodium chloride and other salts in the growth medium are relatively rich in wall teichoic acid, and it was suggested[104] some time ago that these polymers might participate in ion-exchange and the control of access of ions to the inner regions of the cell. This proposal has received powerful confirmation in more recent times. The ability of washed bacterial walls to bind cations, especially bivalent cations, is due to the presence of acidic centres in the wall structure,[105] and it has been shown that the teichoic acid is usually the binding agent.[106] Moreover, there is competition between the cations and the free amino groups of alanine ester residues for the phosphate groups in the wall, the most efficient binding being shown by walls that have had their alanine ester residues removed.

It has been suggested[106] that a major function of both wall and membrane teichoic acids is to maintain a high concentration of bivalent cations in the region of the membrane. This is consistent with the cation binding properties of these polymers, and it is possible that cells are able to control this activity in part by controlling the amount of ester-bound alanine. It explains the structural diversity of these polymers, for many molecules containing phosphate, hydroxyl and amino groups could satisfy the requirements for controlled cation binding. It also explains the vital requirement of cells for membrane teichoic acid; wall teichoic acids would be functionally less important and under certain conditions could be substituted by other acidic polymers. The membrane teichoic acids, being closer to the part of the cell where maintenance of high concentration of cations is required, would have to be structurally more uniform, and in fact they are all polymers of glycerol phosphate and all possess D-alanine ester residues.

In normal conditions of growth the most abundant bivalent cation is Mg^{2+}, and the proposed function of teichoic acids is supported by the observation[100] that organisms grown in a chemostat under conditions of Mg^{2+} limitation of growth possess exceptionally large amounts of wall teichoic acid. Thus the teichoic acid content of the walls is raised in order to scavenge the limited supply of bivalent cations. Similarly, walls from organisms that are grown in the presence of 7·5% sodium chloride contain fewer alanine ester residues in their teichoic acid, presumably in an effort to bind more effectively the bivalent cations in competition with sodium ions.[106]

There are several reasons why it is important that bacterial cells should maintain a high concentration of bivalent cations in the region of the membrane. Many enzymes are bound to the membrane, including those required for the synthesis of membrane, wall and capsular components, e.g. peptidoglycans, teichoic acids, capsular polysaccharides, phospholipids, glycolipids, lipopolysaccharides, etc., as well as those that participate in electron transport and oxidative phosphorylation. The cation requirements of these enymes have not been determined in all cases, but it is known that many require bivalent cations for optimal activity, and in some cases concentrations of 20 mM or even higher are required. The physical integrity of the membrane and its association with ribosomes and mesosomes are also dependent upon relatively high concentrations of Mg^{2+}.

The function of teichoic acids in maintaining the optimum concentration of Mg^{2+} in the region of the membrane can be demonstrated experimentally.[102] Well washed particulate enzyme preparations for the synthesis of wall teichoic acids from *Bacillus licheniformis* do not contain detectable amounts of teichoic acids, and when these preparations are used to study the biosynthesis of teichoic acids it is found that the reaction shows an optimum requirement for Mg^{2+} at about 12 mM; the rate falls off steeply on either side of this value. In a similar study with a wall-membrane preparation in which the membrane fragments still have wall adhering to them, and teichoic acid is demonstrably present, it is found that, provided the preparation has been equilibrated with Mg^{2+} and then well washed before the experiment is commenced, the reaction rate is independent of the concentration of Mg^{2+} in the surrounding medium. Thus the teichoic acid in the wall-membrane preparation acts as a bivalent cation buffer, and provided that it is in the form of its Mg^{2+} salt at the beginning of the experiment the system supplies its own cation requirements and is not dependent upon the outside concentration of these cations. This confirms that an important function of teichoic acids is to provide and maintain a high and controlled cationic environment for optimal activity of membrane-bound enzymes.

C. EFFECT ON CELL DIVISION AND LYSIS

The main function of teichoic acids is probably their role in maintaining the correct cation balance at the membrane. There is evidence, however, that they have additional important roles in individual cases. In *Diplococcus pneumoniae* the dietary requirement for choline arises through the need for this organism to incorporate it into its wall;[107] the choline-containing material in the wall is in fact the ribitol teichoic acid (see p. 43), and it appears that the organism is unable to synthesize choline for this purpose. Choline can be substituted by ethanolamine in the diet and this is incorporated as such into the teichoic acid. This substitution allows the cells to survive and multiply. The new cells however are quite different in their behaviour; they are no longer autolytic, and they grow in the form of long chains containing hundreds of divided but unseparated cells. Thus a small change in the chemical structure of the wall teichoic acid has a profound effect on the ability of the cells to separate. The failure of the new cells to autolyse is probably related to this altered behaviour in cell division.

The possible influence of wall teichoic acid on lytic enzymes in the wall which are believed to be associated with the process of cell division is also illustrated by *Streptococcus zymogenes*. This organism excretes enzymes into the medium which are powerfully lytic towards the walls of a number of other bacteria. The walls of this organism are immune towards its own lytic enzymes, but if the alanine ester residues are removed from the teichoic acid in its walls then autolysis occurs.[108] The nature of the effect of teichoic acids on these lytic enzymes has not been established.

A further secondary function of wall teichoic acids is associated with their overall negative charge. The maintenance of repulsive charges on the surface of unicellular organisms may be a desirable property enabling cell populations to disperse in the medium and thereby utilize nutrients efficiently. It can be shown that removal of teichoic acid from a wall preparation causes it to settle from suspension as a flocculent mass.[106]

VIII. Future Outlook in Research

Future studies on teichoic acids and related acidic polymers in walls and membranes of bacteria can be expected to continue at an undiminished or even accelerated pace. For a considerable time their importance was suspected but not established, and before significant progress could be made on their function it was necessary to determine structures and biosynthetic routes. Although much is now known about both of these, it is likely that new and interesting structural types will be found, and further work will be carried out in establishing the degree of homogeneity of non-uniformly substituted preparations.

The nature of the linkage between peptidoglycan and other wall polymers is by no means fully understood, and a complete understanding of the mechanism of formation of the three-dimensional wall must await clarification of the details of linkage of polymers. This is an exceptionally difficult area of research and progress may not be rapid. The composition of whole walls with respect to the proportion of glycan chains and teichoic acid chains, and variations in composition with changes in growth conditions, will probably receive much attention, as these relate to an understanding of mechanisms of control of cell growth.

One of the most important aspects of the significance of teichoic acids concerns their exact location in both walls and membranes. It is to be hoped that more specific staining techniques might be developed for studies of this subject with the electron microscope; of particular interest will be the relationship between mesosomes and membrane lipoteichoic acids. The general significance of membrane teichoic acids has not received the attention it deserves, yet the work demonstrating the effect of these polymers on cation balance at the membrane illustrates how important they must be in many aspects of membrane behaviour.

ACKNOWLEDGMENTS

The author is indebted to the many colleagues who have worked with him in efforts to understand the nature and activities of bacterial cell walls and membranes. Although it is not possible to mention all by name, a special debt is owed to those who have contributed much over a long period, namely Professor J. G. Buchanan and Drs A. R. Archibald, I. C. Hancock and Helen Hussey. Acknowledgment should also be made to the Medical Research Council and subsequently the Science Research Council for their unfailing support.

REFERENCES

1. Armstrong, J. J., Baddiley, J., Buchanan, J. G., Carss, B. & Greenberg, G. R. (1958). Isolation and structure of ribitol phosphate derivatives (teichoic acids) from bacterial cell walls. *J. Chem. Soc.* 4344-4345.
2. Archibald, A. R., Baddiley, J. & Blumsom, N. L. (1968). The teichoic acids. *Advan. Enzymol. Relat. Areas Mol. Biol.* **30**, 223-253.
3. Sharpe, M. E., Davison, A. L. & Baddiley, J. (1964). Teichoic acids and group antigens in lactobacilli, *J. Gen. Microbiol.* **34**, 333-340.
4. Rogers, H. J. & Perkins, H. R. (1968). *Cell Walls and Membranes* Spon Ltd., London.
5. Hay, J. B., Wicken, A. J. & Baddiley, J. (1963). The location of intracellular teichoic acids. *Biochim. Biophys. Acta,* **71**, 188-190.
6. Shockman, G. D. & Slade, H. D. (1964). The cellular location of the streptococcal group D antigen. *J. Gen. Microbiol.* **37**, 297-305.

7. Rao, E. V., Watson, M. J., Buchanan, J. G. & Baddiley, J. (1969). The type-specific substance from *Pneumococcus* type 29. *Biochem. J.* **111**, 547-556.

8. Anderson, J. C., Archibald, A. R., Baddiley, J., Curtis, M. J. & Davey, N. B. (1969). The action of dilute aqueous *NN*-dimethylhydrazine on bacterial cell walls. *Biochem. J.* **113**, 183-189.

9. Hughes, R. C. & Tanner, P. J. (1968). The action of dilute alkali on some bacterial cell walls. *Biochem. Biophys. Res. Commun.* **33**, 22-28.

10. Archibald, A. R., Baddiley, J. & Heptinstall, S. (1969). The distribution of the glycosyl substituents along the chain of the teichoic acid in walls of *Lactobacillus buchneri* N.C.I.B. 8007. *Biochem. J.* **111**, 245-246.

11. Archibald, A. R., Coapes, H. E. & Stafford, G. H. (1969). The action of dilute alkali on bacterial cell walls. *Biochem. J.* **113**, 899-900.

12. Button, D., Archibald, A. R. & Baddiley, J. (1966). The linkage between teichoic acid and glycosaminopeptide in the walls of a strain of *Staphylococcus lactis*. *Biochem. J.* **99**, 11-14C.

13. Baddiley, J., Buchanan, J. G. RajBhandary, U. L. & Sanderson, A. R. (1962). Teichoic acid from the walls of *Staphylococcus aureus* H. 1. Structure of the *N*-acetylglucosaminylribitol residues. *Biochem. J.* **82**, 439-448.

14. Baddiley, J. Buchanan, J. G., Martin, R. O. & RajBhandary, U. L. (1962). Teichoic acids from the walls of *Staphylococcus aureus* H. 2. Location of phosphate and alanine residues. *Biochem. J.* **85**, 49-56.

15. Davison, A. L. & Baddiley, J. (1963). The distribution of teichoic acids in staphylococci. *J. Gen. Microbiol.* **32**, 271-276.

16. Archibald, A. R. & Baddiley, J. (1966). The teichoic acids. *Advan. Carbohydrate Chem.* **21**, 323-375.

17. Mirelman, D., Beck, B. D. & Shaw, D. R. D. (1970). The location of the D-alanyl ester in the ribitol teichoic acid of *Staphylococcus aureus*. *Biochem. Biophys. Res. Commun.* **39**, 712-717.

18. Torii, M., Kabat, E. A. & Bezer, A. E. (1964). Separation of teichoic acid of *Staphylococcus aureus* into two immunologically distinct specific polysaccharides with α- and β-*N*-acetylglucosaminyl linkages respectively: antigenicity of teichoic acids in man. *J. Exp. Med.* **120**, 13-29.

19. Armstrong, J. J., Baddiley, J. & Buchanan, J. G. (1961). Further studies on the teichoic acid from *Bacillus subtilis* walls. *Biochem. J.* **80**, 254-261.

20. Archibald, A. R., Baddiley, J. & Buchanan, J. G. (1961). The ribitol teichoic acid from *Lactobacillus arabinosus* walls: isolation and structure of ribitol glucosides. *Biochem. J.* **81**, 124-134.

21. Chittenden, G. J. F., Roberts, W. K., Buchanan, J. G. & Baddiley, J. (1968). The specific substance from *Pneumococcus* type 34 (41). The phosphodiester linkages. *Biochem. J.* **109**, 597-602.

22. Rao, E. V., Buchanan, J. G. & Baddiley, J. (1966). The type-specific substance from *Pneumococcus* type 10A(34). The phosphodiester linkages. *Biochem. J.* **100**, 811-814.

23. Brundish, D. E. & Baddiley, J. (1968). Pneumococcal C-substance, a ribitol teichoic acid containing choline phosphate. *Biochem. J.* **110**, 573-582.

24. Kelemen, M. V. & Baddiley, J. (1961). Structure of the intracellular glycerol teichoic acid from *Lactobacillus casei* A. T. C. C. 7469. *Biochem. J.* **80**, 246-254.

25. Baddiley, J. & Davison, A. L. (1961). The occurrence and location of teichoic acids in lactobacilli. *J. Gen. Microbiol.* **24**, 295-299.
26. Archibald, A. R. Baddiley, J. & Button, D. (1968). The membrane teichoic acid of *Staphylococcus lactis* 13. *Biochem. J.* **110**, 559-563.
27. Critchley, P., Archibald, A. R. & Baddiley, J. (1962). The intracellular teichoic acid from *Lactobacillus arabinosus* 17-5. *Biochem. J.* **85**, 420-431.
28. Shabarova, Z. A., Hughes, N. A. & Baddiley, J. (1962). The influence of adjacent phosphate and hydroxyl groups on amino acid esters. *Biochem. J.* **83**, 216-219.
29. Wicken, A. J., Elliott, S. D. & Baddiley, J. (1963). The identity of streptococcal group D antigen with teichoic acid. *J. Gen. Microbiol.* **31**, 231-239.
30. Wicken, A. J. & Baddiley, J. (1963). Structure of intracellular teichoic acids from group D streptococci. *Biochem. J.* **87**, 54-62.
31. Toon, P., Brown, P. E. & Baddiley, J. (1972). The lipid-teichoic acid complex in the cytoplasmic membrane of *Streptococcus faecalis* N.C.I.B. 8191. *Biochem. J.* **127**, 399-409.
32. Ellwood, D. C., Kelemen, M. V. & Baddiley, J. (1963). The glycerol teichoic acid from the walls of *Staphylococcus albus* N.T.C.C. 7944. *Biochem. J.* **86**, 213-225.
33. Glaser, L. & Burger, M. M. (1964). The synthesis of teichoic acids. III. Glucosylation of polyglycerophosphate. *J. Biol. Chem.* **239**, 3187-3191.
34. Hughes, R. C. (1965). The isolation of structural components present in the cell wall of *Bacillus licheniformis* N.C.T.C. 6346. *Biochem. J.* **96**, 700-709.
35. Chin, T., Burger, M. M. & Glaser, L. (1966). Synthesis of teichoic acids. VI. The formation of multiple wall polymers in *Bacillus subtilis* W23. *Archs. Biochem. Biophys.* **116**, 358-367.
36. Shaw, N. & Baddiley, J. (1964). The teichoic acid from the walls of *Lactobacillus buchneri* N.C.I.B. 8007. *Biochem. J.* **93**, 317-321.
37. Archibald, A. R., Baddiley, J. & Heptinstall, S. (1969). The distribution of the glucosyl substituents along the chain of the teichoic acid in walls of *Lactobacillus buchneri* N.C.I.B. 8007. *Biochem. J.* **111**, 245-246.
38. Burger, M. M. & Glaser, L. (1966). The synthesis of teichoic acids. V. Polyglucosylglycerol phosphate and polygalactosylglycerol phosphate. *J. Biol. Chem.* **241**, 494-506.
39. Adams, J. B., Archibald, A. R., Baddiley, J., Coapes, H. E. & Davison, A. L. (1969). Teichoic acids possessing phosphate-sugar linkages in strains of *Lactobacillus plantarum. Biochem. J.* **113**, 191-193.
40. Archibald, A. R. & Coapes, H. E. (1971). The wall teichoic acids of *Lactobacillus plantarum* N.I.R.D. C106. Location of the phosphodiester groups and separation of the chains. *Biochem. J.* **124**, 449-460.
41. Naumova, I. B. & Zaretskaya, M. Z. (1964). Some properties of glyceroteichoic acids from *Streptomyces rimosus* and *Streptomyces antibioticus. Dokl. Akad. Nauk. SSSR*, **157**, 207-210.
42. Wicken, A. J. (1966). The glycerol teichoic acid from the cell wall of *Bacillus stearothermophilus* B. 65. *Biochem. J.* **99**, 108-116.
43. Estrada-Parra, S., Rebers, P. A. & Heidelberger, M. (1962). The specific polysaccharide of type 18 *Pneumococcus*. II. *Biochemistry*, **1**, 1175-1177.

44. Estrada-Parra, S. & Heidelberger, M. (1963). The specific polysaccharide of type 18 *Pneumococcus*. III. *Biochemistry*, **2**, 1288-1294.

45. Heidelberger, M., Estrada-Parra, S. & Brown, R. (1964). The specific polysaccharide of type 18A *Pneumococcus*. *Biochemistry*, **3**, 1548-1550.

46. Kennedy, D. A., Buchanan, J. G. & Baddiley, J. (1969). The type-specific substance from *Pneumococcus* type 11A (43). *Biochem. J.* **115**, 37-45.

47. Archibald, A. R., Baddiley, J. & Button, D. (1968). The glycerol teichoic acid of walls of *Staphylococcus lactis* I3. *Biochem. J.* **110**, 543-557.

48. Archibald, A. R., Baddiley, J., Heckels, J. E. & Heptinstall, S. (1971). Further studies on the glycerol teichoic acid of walls of *Staphylococcus lactis* I3. Location of the phosphodiester groups and their susceptibility to hydrolysis with alkali. *Biochem. J.* **125**, 353-359.

49. Archibald, A. R. & Heptinstall, S. (1971). The teichoic acids of *Micrococcus* sp. 24. *Biochem. J.* **125**, 361-363.

50. Archibald, A. R., Baddiley, J., Button, D., Heptinstall, S. & Stafford, G. H. (1968). Occurrence of polymers containing N-acetylglucosamine 1-phosphate in bacterial walls. *Nature (London)*, **219**, 855-856.

51. Partridge, M. D., Davison, A. L. & Baddiley, J. (1971). A polymer of glucose and N-acetylgalactosamine 1-phosphate in the wall of *Micrococcus* sp. A1. *Biochem. J.* **121**, 695-700.

52. Strominger, J. L. & Ghuysen, J. M. (1963). Linkage between teichoic acid and the glycopeptide in the cell wall of *Staphylococcus aureus*. *Biochem. Biophys. Res. Commun.* **12**, 418-424.

53. Ghuysen, J. M., Tipper, D. J. & Strominger, J. L. (1965). Structure of the cell wall of *Staphylococcus aureus*, strain Copenhagen. IV. The teichoic acid-glycopeptide complex. *Biochemistry*, **4**, 474-485.

54. Knox, K. W. & Hall, E. A. (1965). The linkage between the polysaccharide and mucopeptide components of the cell wall of *Lactobacillus casei*. *Biochem. J.* **96**, 302-309.

55. Hall, E. A. & Knox, K. W. (1965). Properties of the polysaccharide and mucopeptide components of the cell wall of *Lactobacillus casei*. *Biochem. J.* **96**, 310-318.

56. Liu, T. Y. & Gotschlich, E. G. (1963). The chemical composition of pneumococcal C-polysaccharide. *J. Biol. Chem.* **238**, 1928-1934.

57. Burger, M. M. & Glaser, L. (1964). The synthesis of teichoic acids. I. Polyglycerolphosphate. *J. Biol. Chem.* **239**, 3168-3177.

58. Wicken, A. J. & Knox, K. W. (1970). Studies on the group F antigen of lactobacilli: isolation of a teichoic acid-lipid complex from *Lactobacillus fermenti* N.C.T.C. 6991. *J. Gen. Microbiol.* **60**, 293-301.

59. Knox, K. W. & Wicken, A. J. (1971). Serological properties of the wall and membrane teichoic acids from *Lactobacillus helveticus* N.C.I.B. 8025. *J. Gen. Microbiol.* **63**, 237-248.

60. Baddiley, J. & Mathias, A. P. (1954). Cytidine nucleotides. Part I. Isolation from *Lactobacillus arabinosus*. *J. Chem. Soc.* 2723-2731.

61. Baddiley, J., Buchanan, J. G., Carss, B., Mathias, A. P. & Sanderson, A. R. (1956). The isolation of cytidine diphosphate glycerol, cytidine diphosphate ribitol and mannitol 1-phosphate from *Lactobacillus arabinosus*. *Biochem. J.* **64**, 599-603.

62. Baddiley, J., Buchanan, J. G., Mathias, A. P. & Sanderson, A. R. (1956). Cytidine diphosphate glycerol. *J. Chem. Soc.* 4186-4190.

63. Baddiley, J., Buchanan, J. G., Carss, B. & Mathias, A. P. (1956). Cytidine diphosphate ribitol from *Lactobacillus arabinosus*. *J. Chem. Soc.* 4583-4588.
64. Baddiley, J., Buchanan, J. G. & Carss, B. (1957). The configuration of the ribitol phosphate residue in cytidine diphosphate ribitol. *J. Chem. Soc.* 1869-1876.
65. Baddiley, J., Buchanan, J. G. & Sanderson, A. R. (1958). Synthesis of cytidine diphosphate glycerol. *J. Chem. Soc.* 3107-3110.
66. Baddiley, J., Buchanan, J. G. & Fawcett, C. P. (1959). Synthesis of cytidine diphosphate ribitol. *J. Chem. Soc.* 2192-2196.
67. Shaw, D. R. D. (1962). Pyrophosphorolysis and enzymic synthesis of cytidine diphosphate glycerol and cytidine diphosphate ribitol. *Biochem. J.* **82**, 297-312.
68. Glaser, L. (1963). Ribitol 5-phosphate dehydrogenase from *Lactobacillus plantarum*. *Biochim. Biophys. Acta*, **67**, 525-530.
69. Glaser, L. (1964). The synthesis of teichoic acids. II. Polyribitol phosphate. *J. Biol. Chem.* **239**, 3178-3186.
70. Ishimoto, N. & Strominger, J. L. (1966). Polyribitol phosphate synthetase of *Staphylococcus aureus*. *J. Biol. Chem.* **241**, 639-650.
71. Nathenson, S. G. & Strominger, J. L. (1963). Enzymatic synthesis of *N*-acetylglucosaminylribitol linkages in teichoic acid from *Staphylococcus aureus*, strain Copenhagen. *J. Biol. Chem.* **238**, 3161-3169.
72. Nathenson, S. G., Ishimoto, N., Anderson, J. S. & Strominger, J. L. (1966). Enzymatic synthesis and immunochemistry of α and β-*N*-acetyl-glucosaminylribitol linkages in teichoic acids from several strains of *Staphylococcus aureus*. *J. Biol. Chem.* **241**, 651-658.
73. Baddiley, J., Blumsom, N. L. & Douglas, L. J. (1968). The biosynthesis of the wall teichoic acid in *Staphylococcus lactis* I3. *Biochem. J.* **110**, 565-571.
74. Brooks, D. & Baddiley, J. (1969). The mechanism of biosynthesis and direction of chain extension of a poly(*N*-acetylglucosamine 1-phosphate) from the walls of *Staphylococcus lactis* N.C.T.C. 2102. *Biochem. J.* **113**, 635-642.
75. Baddiley, J. & Neuhaus, F. C. (1960). The enzymic activation of D-alanine. *Biochem. J.* **75**, 579-587.
76. Reusch, V. M. & Neuhaus, F. C. (1971). D-Alanine: membrane acceptor ligase from *Lactobacillus casei*. *J. Biol. Chem.* **246**, 6136-6143.
77. Higashi, Y., Strominger, J. L. & Sweeley, C. C. (1967). Structure of a lipid intermediate in cell wall peptidoglycan synthesis: a derivative of a C_{55} isoprenoid alcohol. *Proc. Nat. Acad. Sci. U.S.* **57**, 1878-1884.
78. Brooks, D. & Baddiley, J. (1969). A lipid intermediate in the synthesis of a poly(*N*-acetylglucosamine 1-phosphate) from the wall of *Staphylococcus lactis* N.C.T.C. 2102. *Biochem. J.* **115**, 307-314.
79. Douglas, L. J. & Baddiley, J. (1968). A lipid intermediate in the biosynthesis of a teichoic acid. *FEBS Lett.* **1**, 114-116.
80. Hussey, H. & Baddiley, J. (1972). Lipid intermediates in the biosynthesis of the wall teichoic acid in *Staphylococcus lactis* I3. *Biochem. J.* **127**, 39-50.
81. Hancock, I. C. & Baddiley, J. (1972). Biosynthesis of the wall teichoic acid in *Bacillus licheniformis*. *Biochem. J.* **127**, 27-37.

82. Nikaido, K. & Nikaido, H. (1971). Glucosylation of lipopolysaccharide in *Salmonella*: biosynthesis of O-antigen factor 12_2. II. Structure of the lipid intermediate. *J. Biol. Chem.* **246**, 3912-3919.

83. Anderson, R. G., Hussey, H. & Baddiley, J. (1972). The mechanism of wall synthesis in bacteria. The organization of enzymes and isoprenoid phosphates in the membrane. *Biochem. J.* **127**, 11-25.

84. Higashi, Y., Strominger, J. L. & Sweeley, C. C. (1970). Biosynthesis of the peptidoglycan of bacterial cell walls. XXI. Isolation of free C_{55} isoprenoid alcohol and of lipid intermediates in peptidoglycan synthesis from *Staphylococcus aureus*. *J. Biol. Chem.* **245**, 3697-3702.

85. Wright, A., Dankert, M., Fennessey, P. & Robbins, P. W. (1967). Characterization of a polyisoprenoid compound functional in O-antigen biosynthesis. *Proc. Nat. Acad. Sci. U.S.* **57**, 1798-1803.

86. Scher, M., Lennarz, W. J. & Sweeley, C. C. (1968). The biosynthesis of mannosyl-1-phosphorylpolisoprenol in *Micrococcus lysodeikticus* and its role in mannan synthesis. *Proc. Nat. Acad. Sci. U.S.* **59**, 1313-1320.

87. Siewert, G. & Strominger, J. L. (1967). Bacitracin: an inhibitor of the dephosphorylation of lipid pyrophosphate, an intermediate in biosynthesis of the peptidoglycan of bacterial cell walls. *Proc. Nat. Acad. Sci. U.S.* **57**, 767-773.

88. Watkinson, R. J., Hussey, H. & Baddiley, J. (1971). Shared lipid phosphate carrier in the biosynthesis of teichoic acid and peptidoglycan. *Nature New Biology,* **229**, 57-59.

89. Robbins, P. W., Bray, D., Dankert, M. & Wright, A. (1967). Direction of chain growth in polysaccharide synthesis. *Science*, **158**, 1536-1542.

90. Kennedy, L. D. & Shaw, D. R. D. (1968). Direction of polyglycerol phosphate chain growth in *Bacillus subtilis*. *Biochem. Biophys. Res. Commun.* **32**, 861-865.

91. Hussey, H., Brooks, D. & Baddiley, J. (1969). Direction of chain extension during the biosynthesis of teichoic acids in bacterial cell walls. *Nature* (*London*), **221**, 665-666.

92. Higashi, Y., Siewert, G. & Strominger, J. L. (1970). Biosynthesis of the peptidoglycan of bacterial cell walls. XIX. Isoprenoid alcohol phosphokinase. *J. Biol. Chem.* **245**, 3683-3690.

93. Glaser, L., Ionesco, H. & Schaeffer, P. (1966). Teichoic acids as components of a specific phage receptor in *Bacillus subtilis*. *Biochim. Biophys. Acta*, **124**, 415-417.

94. Wolin, M. J., Archibald, A. R. & Baddiley, J. (1966). Changes in wall teichoic acid resulting from mutations in *Staphylococcus aureus*. *Nature* (*London*), **209**, 484-486.

95. Young, F. E. (1967). Requirement of glucosylated teichoic acid for adsorption of phage in *Bacillus subtilis* 168. *Proc. Nat. Acad. Sci. U.S.* **53**, 2377-2384.

96. Murayama, Y., Kotani, S. & Kato, K. (1968). Solubilization of phage receptor substances from cell walls of *Staphylococcus aureus* (strain Copenhagen) by cell wall lytic enzymes. *Biken J.* **11**, 269-291.

97. Coyette, J. & Ghuysen, J. M. (1968). Structure of the cell wall of *Staphylococcus aureus*, strain Copenhagen. IX. Teichoic acid and phage adsorption. *Biochemistry*, **7**, 2385-2389.

98. Chatterjee, A. N. (1969). Use of bacteriophage-resistant mutants to study

the nature of the bacteriophage receptor site cf *Staphylococcus aureus. J. Bact.* **98**, 519-527.

99. Archibald, A. R. & Coapes, H. E. (1971). The influence of growth conditions on the presence of bacteriophage-receptor sites in walls of *Bacillus subtilis* W23. *Biochem. J.* **125**, 667-669.
100. Tempest, D. W., Dicks, J. W. & Ellwood, D. C. (1968). Influence of growth conditions on the concentration of potassium in *Bacillus subtilis* var. *niger* and its possible relationship to cellular ribonucleic acid, teichoic acid and teichuronic acid. *Biochem. J.* **106**, 237-243.
101. Ellwood, D. C. & Tempest, D. W. (1968). The teichoic acids of *Bacillus subtilis* var. *niger* and *Bacillus subtilis* W23 grown in a chemostat. *Biochem. J.* **108**, 40P.
102. Hughes, A. H., Stow, M., Hancock, I. C. & Baddiley, J. (1971). Function of teichoic acids and effect of novobiocin on control of Mg^{2+} at the bacterial membrane. *Nature New Biology*, **229**, 53-55.
103. Stow, M., Starkey, B. J., Hancock, I. C. & Baddiley, J. (1971). Inhibition by chloramphenicol of glucose transfer in teichoic acid biosynthesis. *Nature New Biology*, **229**, 56-57.
104. Archibald, A. R., Armstrong, J. J., Baddiley, J. & Hay, J. B. (1961). Teichoic acids and the structure of bacterial walls. *Nature (London)*, **191**, 570-572.
105. Meers, J. L. & Tempest, D. W. (1970). The influence of growth-limiting substrate and medium NaCl concentration on the synthesis of magnesium-binding sites in the walls of *Bacillus subtilis* var. *niger. J. Gen. Microbiol.* **63**, 325-331.
106. Heptinstall, S., Archibald, A. R. & Baddiley, J. (1970). Teichoic acids and membrane function in bacteria. *Nature (London)*, **225**, 519-521.
107. Tomasz, A. (1967). Choline in the cell wall of a bacterium: novel type of polymer-linked choline in *Pneumococcus. Science,* **157**, 694-697.
108. Davie, J. M. & Brock, T. D. (1966). Effect of teichoic acid on resistance to the membrane-lytic agent of *Streptococcus zymogenes. J. Bacteriol.* **92**, 1623-1631.

The Application of Magnetic Resonance Methods to the Study of Enzyme Structure and Action

P. F. KNOWLES

*Astbury Department of Biophysics, University of Leeds,
Leeds LS2 9JT, England*

I. Introduction

Magnetic Resonance Spectroscopy has now established itself as a branch of science of particular value to the study of enzyme structure and function. Nuclear Magnetic Resonance Spectroscopy (n.m.r.) and Electron Paramagnetic Resonance Spectroscopy (e.p.r.) both give information on enzyme structure and the dynamics of enzyme action; it is this latter information which perhaps offers the major advantage of magnetic resonance methods over that of X-ray diffraction. However, had the structures of several enzymes in the crystalline state not been known through X-ray diffraction studies, it is doubtful whether the application of magnetic resonance methods to the study of enzymes would have made such rapid progress over the past ten years and it may be anticipated that the two techniques will continue to be complementary. When n.m.r reaches a state of refinement whereby the complete structure of an enzyme in solution may be determined, comparison with the X-ray data should give an answer to the question "Is the structure of an enzyme in the crystalline state identical to that in solution?"

N.m.r. and e.p.r. are themselves complementary in the information they can provide about enzymes; the studies made by Cohn, Mildvan and their coworkers[1] on enzymes which require metals as cofactors show the value of applying both techniques where possible. The differences between n.m.r. and e.p.r. stem from their underlying physical origins.* N.m.r. is a form of spectroscopy based on the absorption of radiofrequency ($\nu \sim 10^7 - 10^8$ Hz) electromagnetic radiation by atomic nuclei having nuclear spin $I > 0$ (most commonly protons where $I = \frac{1}{2}$) when placed in a strong ($\sim 10^4$ oersted) magnetic field. The spectrum obtained (Fig. 1a) when radiofrequency absorption is plotted against radiofrequency (R_f) for a constant magnetic field may be considered analogous to the absorption of ultraviolet or visible radiation by different chemical groups in a molecule as the wavelength is scanned. There is a

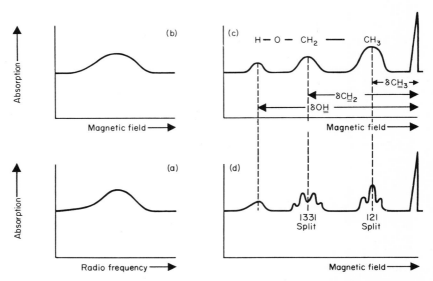

Fig. 1. N.m.r. spectra of ethanol with increasing resolution. (a) Low resolution with R_f sweep, magnetic field constant. (b) Low resolution with magnetic field sweep, R_f constant. (c) Medium resolution. The chemical shifts (δ) relative to a proton standard for the three species of protons are shown. (d) High resolution, showing hyperfine splitting of the proton peaks.

* It is intended that this essay should contain sufficient Magnetic Resonance theory to make the section dealing with enzyme structure and function (Section III) intelligible without recourse to any other review or book.

relationship between the R_f frequency (ν) and the magnetic field (H) in n.m.r. directly comparable to e.p.r. resonance condition in Equation 1, given below. Practically it is found more convenient to scan the magnetic field with a constant radiofrequency; this is shown in Fig. 1b.

A. NUCLEAR MAGNETIC RESONANCE SPECTRA

Three characteristic parameters may be measured from an n.m.r. spectrum and discrete information derived from each:

(1) The *position* of a particular absorption, usually given as a *chemical shift* (δ) relative to a standard, reflects the electron distribution about the nucleus (Fig. 1c); it is the electron distribution which determines the chemistry of a particular atom. In many cases, the spectrum of nuclei in a single chemical group is observed to be a series of absorption lines (Fig. 1d) as a result of "coupling" to neighbouring nuclei. There are ($n + 1$) of such *"hyperfine"* lines where n is the number of identical nuclei on the neighbouring group; the intensities are given by a binomial relationship.

$$
\begin{array}{ccccccc}
 & & & 1 & & & \\
 & & 1 & 2 & 1 & & \\
 & 1 & 3 & 3 & 1 & & \\
1 & 4 & 6 & 4 & 1 & & \\
\end{array}
$$

for one, two, three and four protons respectively. Thus for our example of ethanol, the methylene resonance will be split into a quartet by the three identical protons on the methyl while the methyl resonance will be split into a triplet by the two identical protons on the methylene. It is difficult to see structure on the OH protons since trace amounts of water mask the effect. The complexity to the spectrum from hyperfine interactions contains additional structural information and is the basis for *High Resolution* n.m.r. studies. It should, perhaps, be pointed out that all high resolution n.m.r. studies are made on solutions.

(2) The *intensity* of an absorption is directly related to the number of nuclei in that particular chemical environment.

(3) The *width* of an absorption line (measured in units of frequency at half peak height) is ultimately determined by the rate of atomic motions, and is related to "relaxation processes" which determine how quickly the higher energy spin state can dissipate its energy. From line widths and *relaxation times* (see Section II) the mobility of the nuclei and their exchange between different chemical environments can be derived.

Thus in theory at least n.m.r. can give us complete information on the structure and mode of action of the whole enzyme molecule.

B. ELECTRON PARAMAGNETIC RESONANCE SPECTRA

E.p.r on the other hand is much more specific. Only systems having unpaired electrons, that is *paramagnetic systems* (for example free radicals, some of which are short lived, and certain valence states of transition metal ions), give e.p.r. spectra, which arise from the absorption of microwave frequency ($\nu \sim 10^{10}$ Hz) electromagnetic radiation by electrons (spin $S = \frac{1}{2}$) when placed in a strong magnetic field ($\sim 10^3$ oersted). The electron is localized on a particular atom or at most spread over the atoms adjacent to this particular atom and thus the e.p.r. spectrum reports on this localized part of the enzyme. For example, since in enzymes requiring a transition metal (either strongly bound or dissociable) the metal is directly involved in the catalytic process, e.p.r. "opens a window" through which changes at the active site can be studied *specifically* without any complication of the spectrum by the diamagnetic remainder of the protein. The three characteristic parameters discussed for n.m.r. have direct parallels in e.p.r.

(1) The *position* of an absorption band is related to the chemistry of the paramagnetic site. The equation describing the resonance condition is

$$h\nu = \bar{g}\beta\bar{H}$$

Equation 1

where h is Planck's constant, ν is the microwave frequency which in practice is maintained constant whilst the magnetic field, $\bar{H}$, is scanned, and β is a constant named the Bohr Magneton. The spectroscopic splitting factor, $\bar{g}$, determines the magnetic field value at which resonance occurs. For an unpaired electron spinning about its axis in the free state, $\bar{g} \approx 2 \cdot 00$ but considerable deviations from this value can occur when the electron is constrained by non-spherically symmetric orbital motion. The $\bar{g}$ value varies away from $2 \cdot 00$ as the extent of "spin-orbit coupling" increases. In addition, due to constraint of ligand binding on the spatial geometry of p and d orbitals, the $\bar{g}$ value will depend on the direction along which the magnetic field is applied. For a completely asymmetric environment there will be *three* discrete $\bar{g}$ values depending upon whether the magnetic field is applied along the x, y or z axes. The system is said to show "$\bar{g}$ value anisotropy". Hence information can be obtained from the $\bar{g}$ value on both the identity of the paramagnetic species as well as the geometry of its chemical environment. It will be noted from Fig. 2(a) that the first derivative of the absorption is plotted rather than the absorption itself since this is the form in which most e.p.r. spectra are recorded in practice.

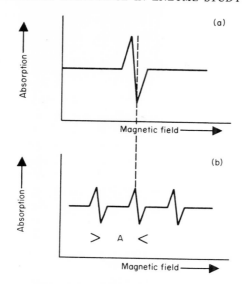

Fig. 2. Single crystal e.p.r. spectra with magnetic field applied along x axis (say). (a) Without nuclear hyperfine splitting. (b) With hyperfine splitting due to coupling with a nucleus spin $I = 1$. The hyperfine splitting constant (A) is shown.

Hyperfine Structure

"Coupling" of the electron magnetic moment with a nuclear moment (either on the same atom or on an adjacent atom) gives rise to $2I + 1$ equally spaced absorption bands for each nuclear moment I. Fig. 2b shows the interaction with a nuclear spin $\bar{I} = 1$. The "hyperfine splitting constant" $(\bar{A})$ is the separation (usually expressed in Hz) between each of these bands and gives added information about the chemistry and symmetry of the paramagnetic centre. For example, the observation of ligand nuclear hyperfine structure in the e.p.r. spectrum of a metal complex implies that the unpaired electron from the metal is delocalized over ligand nuclei, i.e. it implies *covalency*.

(2) The *area* under an absorption band is to a first approximation proportional to the concentration of the paramagnetic species.

(3) On the basis of the resonance equation (Eq. 1) one might expect an e.p.r. absorption to be a sharp line. In practice, however, the lines are broadened due to *relaxation processes* which will be discussed more fully in Section II.

To complete this introduction, perhaps it would be useful to enumerate some of the attractive features of magnetic resonance spectroscopy applied to enzymes. Firstly, measurements can be made in solution at normal temperatures. Secondly, due to the low energy of radio and microwave frequency radiation, the measurement itself does not noticeably perturb the system; nor is the sample destroyed. Finally some information cn be obtained on large enzymes (mol. wt ~ 100,000), on multi-enzyme complexes and on enzymes bound to membranes.

II. Basic Theory

A. WHY THIS SECTION IS NECESSARY

Until recently, an adequate treatment of magnetic resonance theory to satisfy the requirements of organic and biochemists in studying the structure of both small molecules and enzymes could be expressed as "flips" of spin orientation of the magnetic species (for the remainder of this section "magnetic species" means *either* nucleus *or* electron) from alignment *with* the applied magnetic field or *against* the applied magnetic field when a suitable quantum of electronmagnetic radiation is applied (Fig. 3).

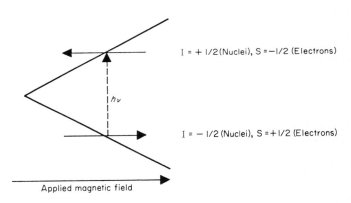

Fig. 3. Simple picture of magnetic resonance in terms of changes in spin orientation.

Now, however, the development of Pulsed and Fourier Transform n.m.r. methods and the growing realization that these "second generation" spectrometers will replace the conventional continuous magnetic or radiofrequency field sweep n.m.r. methods, make it necessary to go rather more fully into the physical processes underlying magnetic resonance, in order that the application of these techniques to the study of enzymes can be understood. The treatment of theory which follows has deliberately been both abbreviated and made qualitative. More complete discussion can be found in a review by Sheard and Bradbury[2] and in the books by Lynden Bell and Harris[3] and by Farrar and Becker.[4]

B. THE MEANING OF "RESONANCE" SPECTROSCOPY

In the Introduction it was stated that the most frequently encountered nuclei in n.m.r. have nuclear spin $\bar{I} = \frac{1}{2}$ whilst an unpaired electron also has a spin, $\bar{S} = \frac{1}{2}$. Quantum Theory only allows $\bar{I} = \frac{1}{2}$ or $\bar{S} = \frac{1}{2}$ to have two orientations parallel or antiparallel with respect to the applied field instead of a continuous distribution of orientations. A useful "classical" understanding of the system can be achieved by considering the magnetic species spinning about its own axis and thereby generating a magnetic moment. When a magnetic field $\bar{H}_0$ is applied externally in a particular direction (defined as the z axis) the particle's magnetic moments, corresponding to $\bar{I}$ or $\bar{S} = \pm \frac{1}{2}$, attempt to line up along this direction but, due to their spin, they in fact are forced to "precess" about the $\bar{H}_0$ field with a frequency ν_0, the *Larmor Frequency*. The analogy to a gyroscope precessing at an angle to the earth's gravitational field is reasonable. When an assembly of particles is being considered it is easiest to draw them with the tails of each vector at a common origin (Fig. 4).

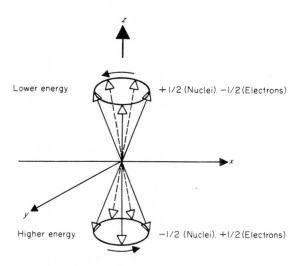

Fig. 4. Precession of magnetic moments about a magnetic field (H_0) applied along the z axis.

When an alternating (R_f) field (either radio or microwave frequency depending upon whether we are considering nuclei or electrons respectively) is applied in the xy plane, the moments of the charged particles attempt to line up *with* this field as well as the $\bar{H}_0$ magnetic field. At resonance we have a situation as in Fig. 5 where rotation about (say) the x axis is occurring simultaneously with rotation about the Z axis. Resonance occurs *only* when the R_f frequency is identical to the Larmor Frequency, ν_0.

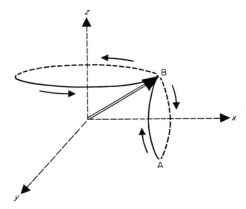

Fig. 5. Resonance condition for magnetic field applied along the z axis, R_f field applied in the xy plane.

During the period A → B, energy is *absorbed from* the alternating field by the charged particle whilst in the period B → A, energy is *given to* the alternating field by the charged particle. The reason why we see a nett *absorption* of radiation during resonance is because the population of spins in the lower energy condition (n_{lower}) is greater than the population in the upper energy condition (n_{upper}). The Boltzmann equation $n_{upper} = (n_{lower}) e^{-\Delta E/kT}$ relates these two populations to the energy difference between them (ΔE) and absolute temperature (T). From this equation it follows that the population difference will be larger when the value of ΔE is high and the value of T is low. For electrons at 298°K in a magnetic field of 3000 gauss $n_{upper}/n_{lower} = 0.9986$ whilst for nuclei at 298°K in a magnetic field of 10,000 gauss, $n_{upper}/n_{lower} = 0.999997$.

Two facts emerge from these figures:

(a) Very sensitive techniques must be used to measure the absorption of radiation, particularly in the n.m.r. case, since the populations in the two energy levels are extremely close. The population difference can be increased by the use of higher magnetic fields.

(b) If too much radiation is applied to the system, the populations of the two energy levels will become equal. There will then be no preference for absorption of radiation over emission and the resonance signal becomes *saturated*. Relaxation processes relieve this saturation condition—the more efficient the relaxation process, the less likely is the signal to become saturated.

C. RELAXATION PROCESSES

The ultimate width of any spectral line is given by Heisenberg's Uncertainty Principle.

$\Delta E. \Delta t = \hbar$ where ΔE is the spectral line width, Δt is the life time of the upper spin state and $\hbar$ is Planck's constant h divided by 2π. Thus the longer the lifetime of the upper spin state, the sharper the lines will be. The value of Δt for nuclei is frequently much greater than for unpaired electrons (particularly in transition ions) and consequently n.m.r. lines are sharper. The width of e.p.r. lines can be reduced by decreasing the temperature (which effectively increases Δt) and many e.p.r. studies on metalloenzymes are therefore carried out at liquid nitrogen temperatures.

Relaxation processes govern the lifetime of the upper spin state. Relaxation is due to interaction between the charged species in the upper spin state and fluctuating magnetic fields originating from any other dipole in the system. Each relaxation process has an associated *Correlation Time*. In general, the correlation time (τ_c) is the average time between molecular collisions for molecules in some state of motion. It is extremely useful to consider the reciprocal of the correlation time ($1/\tau_c$) as the sum of terms, one of which may be large compared with the other terms in a particular situation. Thus

$$\frac{1}{\tau_c} = \frac{1}{\tau_R} + \frac{1}{\tau_M} + \frac{1}{\tau_S} \qquad \text{Equation 2.}$$

where $\dfrac{1}{\tau_R}$ is the rate at which the molecule is tumbling (τ_R is called the rotational correlation time),

$\dfrac{1}{\tau_M}$ is the rate at which the two dipoles approach each other (τ_M is called the "residence" time which is a measure of how long the two dipoles are close together) and

$\dfrac{1}{\tau_S}$ is the electron spin lattice relaxation rate (τ_S is the time taken for the upper electron spin state to dissipate its excess energy).

Correlation times are very important in both n.m.r. and e.p.r.

D. CORRELATION TIMES AND MOLECULAR MOTION

1. *N.m.r.–Limiting Size of Enzymes Whose Structure Can be Studied by N.m.r.*

When a molecule is tumbling in solution at a rate ($1/\tau_R$) comparable to the frequency of spectral line separation, n.m.r. hyperfine lines cannot be resolved completely. As r the radius increases, the correlation time ($\tau_c \approx \tau_R$) *increases* (see Equation 3, below) and $1/\tau_R$ *decreases*. This accounts for a practical upper limit of about ~ 20,000 mol. wt in obtaining *complete* hyperfine structural information on a protein molecule in solution. Nevertheless, useful n.m.r. information can still be obtained on, for example, inhibitor binding to enzymes since the inhibitor might still have considerable mobility even when bound to a large enzyme. The studies made by Raftery and co-workers on the binding of substrate analogues to lysozyme (EC 3.2.1.17) are a good example of this[5] (see Section IIIB).

2. *E.p.r-Spin Labelling*

Throughout this section, the *dynamics* of the resonance process have been emphasized. The technique of "Spin Labelling" (which has made enzymes and other biological molecules not possessing an inherent paramagnetism amenable to study by e.p.r) is able to give very useful information on freedom of movement in specific regions of the enzyme. Spin labels[6] are stable free radicals, the most commonly used having the general formulae

There is an unpaired electron on the nitrogen atom which, through coupling with the ^{14}N nuclear spin ($I = 1$), produces a three line spectrum (Fig. 6A). The grouping R on the molecule can be varied widely by chemical modification; a range of reagents for spin labelling specific sites on enzymes or substrates is therefore already available and its extension awaits the ingenuity of organic chemists.

The Hyperfine Splitting Constant ($\bar{A}$) of a spin label spectrum shows anisotropy in a similar way to the $\bar{g}$ value anisotropy discussed in the Introductory section. Thus the *orientation* of an attached spin label with respect to the axes of an enzyme molecule can be determined.

It is instructive to observe the effect of increasing viscosity in the medium on a spin label spectrum.

The correlation time, τ, may be considered here to be a measure of how rapidly the molecule is tumbling and is related to the viscosity by the equation

$$\tau = \frac{4\pi\eta r^3}{3\,kT}$$
Equation 3.

where η is the viscosity, r is the radius of the spin labelled molecule and T is the absolute temperature.

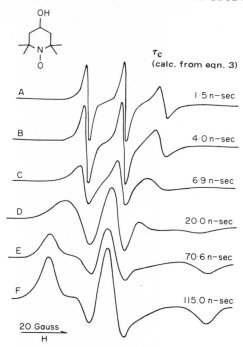

Fig. 6. E.p.r. spectra (run at $0°C$) of 1-oxyl-2,2,6,6-tetramethyl-4-piperidinol in aqueous solutions (A to F) of increasing viscosity. (Fig. by courtesy of Dr I. C. P. Smith.[6a])

As the viscosity is increased the correlation time increases. When the correlation time has increased to $\sim 10^{-8}$ second there is very appreciable distortion of the spectrum (Fig. 6). The reason for this can be considered to be due in part to the fact that the molecule is now tumbling at a rate comparable to the frequency separation of the hyperfine lines. A more detailed consideration of the changes in the spectra is beyond the scope of this essay and is in any event incompletely understood.[6] However, it can be seen, qualitatively, how spin labelling studies can provide information about mobility at specific sites in an enzyme.

E. SPIN HAMILTONIANS

Most research reports on magnetic resonance, particularly those concerning e.p.r., refer to "Hamiltonians". This, from the viewpoint of pure physics, is the most satisfactory way of discussing magnetic resonance phenomena. A simple introduction to the subject here may show how the experimentally determined parameters, for example $\bar{g}$ and $\bar{A}$ values in e.p.r., can be used to provide information on atomic and molecular orbitals.

The Hamiltonian offers a "bridge" between experiment and theory. The basic equation of wave mechanics is $\mathcal{H}\psi = E\psi$ which is a statement of the principle of energy conservation. The left-hand side of the equation is an "operation" upon a wave function (ψ) and the operator is termed the "Hamiltonian" ($\mathcal{H}$). If we can obtain an expression for the Hamiltonian operator, we can derive information on the energy states (E) of the system.

The Quantum Mechanical approach is to write down the complete Hamiltonian for an atom or molecule. For magnetic resonance phenomena, we are considering low energy terms only and a "Spin Hamiltonian" ($\mathcal{H}_{spin}$) is adequate to explain the observed spectra.

$$\mathcal{H} \text{ spin} = \beta_e \, \bar{S} \cdot \bar{g}_e \cdot \bar{H} + \bar{S} \cdot \bar{A} \cdot \bar{I} + \beta_N \bar{I} \cdot \bar{g}_N \cdot \bar{H} \qquad \text{Equation 4.}$$

Succeeding terms in equation 4 have decreasing energy. The first term gives the interaction between the electron spin $\bar{S}$ and the magnetic field $\bar{H}$; $\bar{g}_e$ is a second rank tensor introduced already as the "spectroscopic splitting factor". A question which most biochemists will ask is "What are tensors?". In answer, one must first consider vectors. Vectors, written as "$\bar{X}$" in equation 4, are parameters with a magnitude and a *single* direction; whilst properties of matter like stress, for example, are associated with *two* directions.

Vectors are tensors of the *first* rank. *Stress* and the *spectroscopic splitting factor* are both tensors of the *second* rank.

The second term in the Spin Hamiltonian gives the interaction between the electron spin $\bar{S}$ and the nuclear spin $\bar{I}$. The hyperfine splitting constant, $\bar{A}$, is again a *second* rank tensor. From this second term we can see that $\bar{A}$ is independent of the magnetic field and the use of different magnetic fields offers a way to distinguish between lines in an e.p.r. spectrum originating from differing $\bar{g}$ values as opposed to differing $\bar{A}$ values. The third term gives the interaction between the magnetic field, $\bar{H}$, and the nuclear spin $\bar{I}$, and is the highest energy term (giving rise to chemical shifts) when we are considering an n.m.r. spectrum. The energy difference between terms one and three in the spin Hamiltonian is approximately 1000 times, which explains why e.p.r. is a much more *sensitive* technique than n.m.r. A further term could be added giving the interaction between two different nuclear spins; this would "explain" nuclear hyperfine lines.

F. LONGITUDINAL AND TRANSVERSE RELAXATION

We now need to think of relaxation processes in terms of precession of moments as in Figs 4 and 5. The application of a *pulse* of R_F field (H_1) at the resonance frequency along the x axis will induce a resultant *magnetization* in the

x direction which has two effects:

(1) It causes the spins to precess about the direction of $\overline{H}_1$,
and (2) It aligns all the randomized spins in the xy plane along the x axis.

The relaxation times (defined generally as the times taken for systems to change to a fraction $1/e$ of their original value) for the magnetization effects 1 and 2 to revert to the original state are termed T_1 (longitudinal) and T_2 (transverse) respectively. An approximate value for T_2 in the case of n.m.r. is given by the line width. The situation is illustrated in Fig. 7.

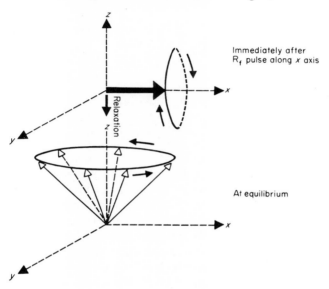

Fig. 7. Relaxation of magnetization following R_f pulse at the Larmor frequency along the x axis. A constant magnetic field is being applied along the z axis.

G. PULSED AND FOURIER TRANSFORM N.M.R.

In simple pulsed n.m.r. a single R_f frequency is applied as a pulse and the decay of the effect (either by longitudinal or transverse relaxation) with respect to time is recorded. In Fourier Transform n.m.r. equally spaced R_f frequencies spanning the whole resonance spectrum are pulsed on to the sample simultaneously and the decay of each effect is recorded in a storage computer. Thus we have the situation in Fig. 8a.

Instead of considering the data in this form, a *Fourier transform* can be performed. By means of this transformation the residual magnetization after a *constant* time t plotted against ν, the R_f frequency, is obtained (Fig. 8b). This

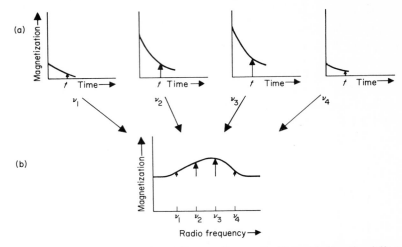

Fig. 8. Fourier Transform N.M.R. (a) Decay of magnetization with time for different R_f frequencies ν_1, ν_2, ν_3, and ν_4. (b) Fourier Transform of these decays to give conventional magnetization against R_f frequency presentation.

produces a conventional R_f absorption *versus* frequency spectrum as in Fig. 1a. As a single statement we can say that "the decay of the signal following an R_f pulse and the absorption spectrum are Fourier transforms of each other". The advantages of such pulsed techniques over continuous scan methods are:

(1) No machine time is wasted as barren regions of the spectra are traversed.

(2) Repeated pulsing and data accumulation allow spectral "noise" to be eliminated. This is particularly important when the nucleus ^{13}C, present only in 1% natural abundance, is being studied. ^{13}C n.m.r. is a major development area of particular value for biological macromolecule studies. The reason for the interest in this field is because the chemical shifts (see Introduction) for ^{13}C are approximately 10 times larger than for protons, and therefore carbon atoms in only slight differing chemical environments can be resolved more readily than protons. One can, of course, increase the resolution on any n.m.r. spectra by increasing the magnetic field but building magnets of the required field stability for operation at fields in excess of say 50,000 oersted presents considerable practical difficulties for the manufacturers and financial problems for the customer.

III. Examples of the Application of Magnetic Resonance Methods to the Study of Enzymes

Emphasis has been placed already on the complementary nature of n.m.r. and e.p.r. Some examples of the application of the two techniques to enzymes will now be given. The examples are intended to be illustrative and in no way comprehensive.

A. STRUCTURAL STUDIES

1. *General Considerations*

It has been pointed out in Section II that high resolution n.m.r. is limited to enzymes of mol. wt < 20,000. Most of the hydrolytic enzymes, the structures of which have been determined in the crystalline state, fall into this category. The most comprehensive studies have been made by Jardetzky[7] and co-workers, who have attempted to establish rules defining the relationship between the spectra

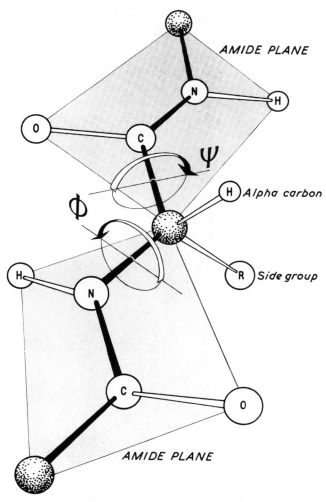

Fig. 9. Rotations about peptide Cα–C and Cα–N single bonds. (Adapted from Dickerson and Geis, ref. 8.)

of the summed component amino acids and of the whole protein. Deviations between these two spectra arise from the secondary and tertiary structure of the protein. Such an analysis would be simplified if certain conformations of the protein could be eliminated and the combined use of conformational analysis and n.m.r. promises to be extremely valuable. A structure for the peptide Gramicidin S (admittedly not an enzyme) has been determined by this approach in which the preferred conformations from rotation about the two single bonds of each amino acid (the ϕ and ψ angles of Ramachandran plots, see Fig. 9 and reference 8) are related to the n.m.r. spectra[9,10]. Computational methods aid in refining the conformational analysis to give the best fit to the n.m.r. data.

The assignment of resonances in the n.m.r. spectrum of an enzyme to particular amino acids is extremely difficult. Jardetzky and his co-workers[11] have succeeded elegantly in the case of the four histidines of ribonuclease (EC 2.7.7.16). Histidine, being a pseudo-aromatic amino acid, has its ring proton n.m.r. resonances shifted down field away from the main body of amino acid resonances (through so-called "ring current" shifts) and it has been possible, by analysis of the n.m.r. spectra at various pH values, to determine the apparent pK for each histidine and thereby to postulate which histidines participate in the

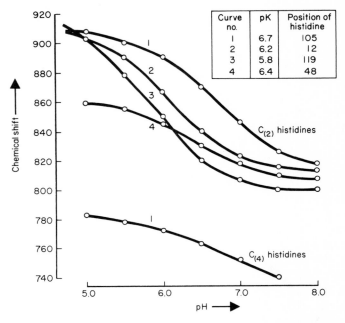

Fig. 10. Titration curves of Histidine-C_2 hydrogen peaks and one Histidine-C_4 hydrogen peak of ribonuclease. (From Jardetzky and Wade-Jardetzky, ref. 7.)

catalytic mechanism, (see Fig. 10). The postulate of histidines 12 and 119 at the active site is in complete agreement with the location of these amino acids in the protein as determined from X-ray diffraction studies.[12] It should be emphasized that n.m.r. is the only method available which allows the assignment of a specific dissociation constant to a specific amino acid.

Selective deuteration of enzymes offers a way to simplify n.m.r. spectra since deuterium has $I = 1$, which has the effect of broadening the resonance signal. Thus by biosynthetic deuteration the enzyme staphylococcal nuclease (EC 3.1.4.7) can be obtained in a state such that all the amino acids except tyrosine and histidine are deuterated.[13] Unambiguous assignment of the resonances from these amino acids then becomes possible. Of course this approach is restricted to enzymes from micro-organisms; an alternative approach has been proposed by Raftery.[14]

An N terminal 20 amino acid peptide can be selectively cleaved from ribonuclease S. This peptide can be fluorinated at lysines 1 and 7 and reattached to the enzyme which recovers full activity. The fluorine ($I = \frac{1}{2}$) n.m.r. spectra can now be studied. To date the technique has been used only to monitor conformational changes in the enzyme following inhibitor binding but it has considerable potential. However, one should always bear in mind that such "nuclear spin labelling" might perturb the enzyme's action. Similar criticisms have been made of the e.p.r. "spin labelling" technique.

An alternative approach to the assignment of n.m.r. resonances to specific amino acids is by comparison of a number of mutant forms of an enzyme; this approach bears analogy to "finger printing" in protein sequence analysis. Studies on mutant forms of cytochrome c have been made.[15] The recent elucidation of the structure of one form of cytochrome c[16] and the known sequences of many other forms[17] will stimulate effort in this field.

Undoubtedly Pulsed Fourier Transform [13]C n.m.r. offers the best *direct* approach to enzyme structural analysis.[18]

2. Haem Proteins

The enzyme structural studies discussed so far have all been made by high resolution n.m.r. Haemoglobin (an "honorary" enzyme) has been extensively investigated by both n.m.r. and e.p.r. Large single crystals of myoglobin and haemoglobin have been available for many years and Ingram[19] made classical e.p.r. studies on the $\bar{g}$ value variation of the haem iron as the single crystals were rotated in the magnetic field.

The haem molecule (Fig. 11) is symmetrical in the plane shown—this is described as "axial symmetry". For e.p.r., therefore, we have one $\bar{g}$ value ($\bar{g}_\perp$) in the plane and one $\bar{g}$ value ($\bar{g}_\parallel$) perpendicular to the plane. The values for haem iron (oxidized from $Fe^{II} \rightarrow Fe^{III}$ for practical convenience) are $\bar{g}_\perp = 6$ and $\bar{g}_\parallel = 2$. Thus for myoglobin with a single haem molecule the orientation of the

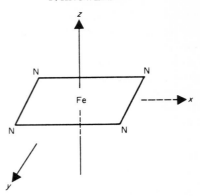

Fig. 11. Geometry of the haem plane(s) in myoglobin and haemoglobin. Pyrrole ring nitrogens are shown as "N".

haem plane with respect to the crystallographic axes can be determined by mounting the crystal along one crystallographic axis and rotating it until a $\bar{g}$ value of 6 is obtained. The magnetic field is then being applied across the haem ring plane. For haemoglobin, the analysis is more complex since there are four haems to consider but Ingram successfully placed each haem with respect to the crystallographic axes. These e.p.r. experiments were carried out *before* the structure of myoglobin and haemoglobin were known and did in fact significantly aid the complete structural analysis.

Spin Labelled Chymotrypsin. Whilst chymotrypsin (EC 3.4.4.5) is not a haem protein, a recent study from McConnell's group[20] extends the orientation studies discussed above. Spin labelled chymotrypsin can be obtained as single crystals by a very neat method: The native enzyme crystals are allowed to react at acid pH with the compound

Under these conditions, a stable acyl spin-labelled form of the enzyme is produced with release of *p*-nitrophenol. The orientation of this rigidly bound spin label with respect to the crystallographic axes can be determined by studying the A value anisotropy as the crystal is rotated in the magnetic field, and information thereby derived concerning the geometry of bound substrate at the active site.

Conformational Changes in Haemoglobin. For several years there has been controversy over the nature of subunit interactions. The Allosteric Model[21] requires that only two conformations in the quaternary structure, termed the R and T forms, exist and that symmetry is preserved during the transition R $\rightleftharpoons$ T. This implies that $\beta - \beta$ and/or $\alpha - \alpha$ subunit interactions which are initiated by binding of an effector molecule (O_2 in the case of haemoglobin) may occur, but *not* $\alpha\beta$ interactions.

The Sequential Model[22] of which the allosteric model is a special case, does *not* require preservation of symmetry and $\alpha\beta$ interactions as well as $\alpha\alpha$ and $\beta\beta$ interactions are permitted. McConnell[23] has elegantly used spin-labelling to distinguish between these possibilities. The following spin-label specifically

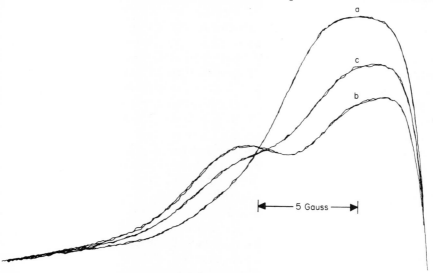

binds at cysteine-93 on the β chains, as shown by X-ray crystallography.[24] The e.p.r. spectra of this spin-labelled haemoglobin in solution at various oxygen concentrations (Fig. 12) fails to show isosbestic points. If these *had* been

Fig. 12. E.p.r. spectra of a 1% human haemoglobin solution labelled with iodoacetamide spin label in 0·1 M-phosphate pH 7·5, 18°C. (a) Deoxy (b) oxy (c) intermediate oxygenation. Each spectrum is recorded twice to indicate the reproducibility. (From McConnell and McFarland, ref. 6.)

observed then this would have been evidence for two and only two conformations of the protein, i.e. these results favour the Koshland rather than the Monod
model for co-operative interactions in haemoglobin.

This finding has been confirmed by studies on (1) (cyanomet α)$_2$ (spin-
labelled β)$_2$ and (2) α_2 (cyanomet spin-labelled β)$_2$. In this terminology, normal
haemoglobin would be written as $(\alpha)_2(\beta)_2$. The cyanomet form of a haemoglobin subunit has a cyanide group on the 6th coordination position and
therefore cannot bind oxygen. Oxygenation of (1) produced a change in the
spectrum like that of simple spin-labelled haemoglobin whilst with (2)
oxygenation of the α chain produced a small but clear change in the e.p.r. of the
spin-labelled β chains. Thus clearly $\alpha\beta$ interactions *must* occur, supporting the
Koshland model.

It is perhaps worth noting that the conformational changes which are being
monitored here by e.p.r. are less than 1 Å and would not be resolved directly by
X-ray methods (1·5 Å limit). This is a further advantage of magnetic resonance
methods over X-ray diffraction.

B. ACTIVE SITE AND SMALL MOLECULE BINDING STUDIES

Both n.m.r. and e.p.r. have been used in studies of the active site. Raftery[5]
has elegantly monitored by n.m.r. the chemical shifts in inhibitor and substrate
protons following binding to lysozyme and has, for example, been able to
quantitate his data to show three subsites associated with binding of chitotriose.
The aggreement with X-ray diffraction data on lysozyme-inhibitor complexes is
good though not complete. He gives a method, which should have wide
applicability for calculating from the chemical shift the dissociation constant of
the inhibitor/enzyme complex.

Similar studies have been made on the binding of sulphonamide inhibitors to
carbonic anhydrase (EC 4.2.2.1). Lanir and Navon[25] and Burgen *et al.*[26] have
both used n.m.r. whilst Coleman *et al.*[27] have used e.p.r. Lanir and Navon used
continuous scan n.m.r. to determine the resonances of protons in the inhibitor
with and without added enzyme; variation of the temperature and R_f frequency
enabled a main conclusion to be drawn that the inhibitor is highly immobilized
following binding to the enzyme; the correlation time τ_c (see Section II)
measured agrees with that calculated for the tumbling of the whole enzyme
molecule. Since the τ_c was unchanged when zinc-free carbonic anhydrase was
used, it could be further concluded that binding probably involved hydrophobic
interactions and *not* attachment to the zinc. The studies reported by Burgen *et
al.* and some of the studies made by the Yale group[27] were on a modified form
of carbonic anhydrase in which the diamagnetic zinc atom bound to the enzyme
had been replaced by the paramagnetic ion Co II; this modified enzyme still
retains its enzymic activity (by contrast the enzyme having Cu II bound is

inactive). It has been emphasized in Section II that the electron dipole is approximately 1000 times that of the nucleus and thus nuclei which are in the proximity of a paramagnetic ion (like the Co II of substituted carbonic anhydrase) will show considerable chemical shifts.

Coleman *et al.*[27] from their studies on the e.p.r. spectra of the cobalt enzyme have speculated that "the transition state (of the enzyme substrate complex) involves a tetrahedral distortion of the coordination geometry". This is in

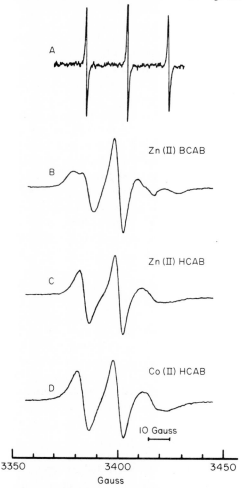

Fig. 13. E.p.r. spectra of 1×10^{-4} M spin-labelled sulphonamide, run at pH 9·0 and 299°K. A. Alone; B. Plus equimolar Zn^{II}Bovine carbonic anhydrase B. C. Plus equimolar Zn^{II}Human carbonic anhydrase B. D. Plus equimolar Co^{II}Human carbonic anhydrase B. (From Taylor, Mushak and Coleman, ref. 27.)

keeping with the predictions of Vallee and Williams[28] that *even when a substrate molecule is not bound*, the metal site of many metallo-enzymes shows distortion which effectively reduced the energy barrier of the transition state. In addition Coleman *et al.* studied the binding of spin-labelled sulphonamide to both zinc and cobalt carbonic anhydrase. They were able to conclude that:

(a) the sulphonamide becomes completely immobilized on binding to the enzyme and any mobility results from rotation of the whole enzyme-inhibitor complex. Addition of cyanide displaces the sulphonamide completely from the enzyme; this can also be seen in the e.p.r. spectrum.

(b) no spin-spin interaction between the Cobalt II and the spin-labelled sulphonamide occurs. This conclusion results from the identical form of spectra C and D as shown in Fig. 13 and indicates that the sulphonamide does *not* bind to the metal or lie in the immediate environment of the metal. The aggreement between the n.m.r. and the e.p.r. results is excellent.

Shulman[29] and his co-workers at Bell Telephones were able to show that the carboxypeptidase A (EC 3.4.4.1) inhibitor, indole acetic acid, binds to the metal. In these studies the native Zn^{2+} ion was replaced by the paramagnetic ion Mn^{2+} and the inhibitor proton resonance lines were studied by field sweep high resolution n.m.r. It was mentioned during the discussion of relaxation processes in Section II that a value for T_2, the transverse relaxation time, can be obtained from the spectral line width. Shulman concludes that T_2 is dominated by the chemical exchange of the inhibitor into the coordination sphere of the manganese; calculations of the inhibitor binding constant, K_i, and the rate constant, k_{off} for the reaction.

$$\text{Enzyme and Inhibitor} \underset{k_{off}}{\overset{k_{on}}{\rightleftharpoons}} \text{Enzyme-Inhibitor}$$

can be made. A value for k_{on} can readily be derived since $K_i = k_{on/off}$. In addition the *distance* of each of the inhibitor protons from the manganese ion can be calculated to an accuracy of approximately 20% and thus some idea of the structural geometry of the metal-inhibitor complex obtained. Recent work from the Oxford Biophysics group[30] shows how paramagnetic probes in conjunction with n.m.r. *and* conformational analysis might be a powerful tool in studying the geometry of enzyme action sites.

C. METALLO-ENZYMES AND METAL ACTIVATED ENZYMES

Approximately 30% of the enzymes listed by Dixon and Webb[31] require a metal for their action. The metal can either be intimately bound (metallo-enzyme) or dissociable. Many metallo-enzymes contain one (or more) transition metal ion(s) and may therefore be conveniently studied by e.p.r. as well as by n.m.r.; one can consider such metallo-enzymes as having their own "built in" spin-label. An excellent review of e.p.r. studies on the metallo-enzymes which are involved in oxidative processes is available.[32]

The studies made by Malmström, Vanngard and co-workers[33] on the copper containing enzyme *laccase* (EC 1.10.3.2) is a good example with which to illustrate the application of e.p.r. to metallo-enzyme structure and function. Laccase catalyses the reaction:

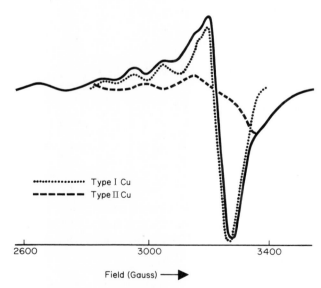

The enzyme contains four atoms of copper per mole of enzyme. Two of these coppers are paramagnetic (Cu^{2+}) and two are diamagnetic (these latter could be Cu^+ *or* spin paired Cu^{2+}).

The e.p.r. spectrum of laccase at 9 GHz microwave frequency is shown in Fig. 14.

Fig. 14. E.p.r. spectra of laccase at 9GHz. Computer simulated, spectrum (solid line), assuming equimolar amounts of *Type I copper* (dotted line) and *Type II copper* (dashed line). The experimentally determined spectrum is identical to the computer simulated (solid line) spectrum. (Based on Malström, Reinhammer & Vänngård, ref. 33.)

The spectrum is determined at liquid nitrogen temperatures on an aqueous solution and shows a structure typical of the polycrystalline state. When the spectrum is recorded at higher frequencies (35 GHz), sufficient resolution is available to allow two different cupric centres (termed Types I and II) to be distinguished. (Fig. 15).

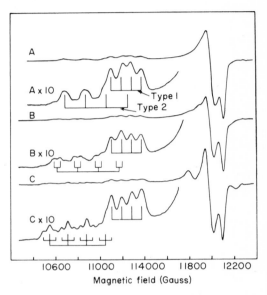

Fig. 15. E.p.r. spectra of native and fluoride inhibited laccase at ~ 35GHz. A. Native enzyme 0·7 mM: the hyperfine structure due to *Types I and II copper* are indicated. B. Native enzyme 0·7 mM plus 0·7 mM sodium fluoride. The hyperfine structure due to interaction between *Type II copper* and *one* fluoride nucleus ($I = \frac{1}{2}$) is indicated. C. Native enzyme (0·7 mM) plus 14 mM sodium fluoride. The hyperfine structure due interaction between *Type II copper* and *two* fluoride nuclei is indicated. (From Malkin, Malmström and Vänngård, ref. 34.)

Type I copper is characterized by having a narrow hyperfine splitting ($A_{\parallel}$) of 250 MHz; compare this value with an $A_{\parallel}$ of approximately 470 MHz for simple copper complexes like copper EDTA. This is an example of the distortion of coordination geometry discussed in Section IVb. The narrow hyperfine splitting indicates extensive delocalization of the electron from the copper which would suggest that Type I copper has a redox function. In addition, Type I copper has an intense blue colour (also an indication of distortion of the metal site), a high redox potential (+ 790 mV) and is readily reduced. Reduction leads to loss of both the blue colour and the e.p.r. signal supporting the view that Type I copper has a redox function.

The other Cu^{2+} (Type II) has an $A_{\parallel}$ of 475 MHz and its e.p.r. signal remains after the addition of a reductant, i.e. it is not reduced to Cu^{+}. Its function, therefore, presents something of a puzzle. Recently[34] it has been found that Type II copper binds inhibitor molecules like cyanide and fluoride; unambiguous evidence for this can be obtained using inhibitors containing atoms with nuclear spin e.g. $K^{13}CN$ and $Na^{19}F$ when the type II copper e.p.r. lines show hyperfine splitting (Fig. 15).

Thus type II copper might function as a site in the enzyme for binding effector molecules which regulate the enzymic activity.

The copper/pyridoxal-containing enzyme *benzylamine oxidase* (EC 1.4.3.4) isolated from pig plasma is under investigation in the Biophysics Department at Leeds. The two cupric atoms in the enzyme, both of which are e.p.r. detectable, are not reduced by addition of the substrate benzylamine, so we have a similar puzzle to that of type II copper in laccase. What part does the copper play in the catalytic mechanism? Changes in the e.p.r. line shape following benzylamine addition suggest that substrate binding to the enzyme affects the copper either directly or, more probably, indirectly. Pulsed n.m.r. is being used as an alternative mode of investigation. Solvent water proton relaxation times are affected appreciably in the presence of the enzyme due to exchange of waters into the Cupric coordination sphere. T_1 for pure water is 3·6 seconds at 24°C which decreases to approximately 200 milliseconds in the presence of 6×10^{-4} M enzyme. A plot of ln T_1 *versus* the reciprocal of the absolute temperature ($\theta °K$) is shown in Fig. 16. The *low* temperature part of the plot shows a positive linear dependence on $1/\theta$ which is interpreted to mean that τ_c, the correlation time for the water proton relaxation, is governed by τ_S, the electron-spin lattice relaxation time. At high temperatures it is proposed that chemical exchange τ_M dominates the relaxation process.

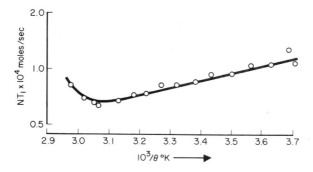

Fig. 16. Dependence of NT_1 (where N is the enzyme concentration in moles, T_1 is the longitudinal relaxation time in seconds) on temperature ($\theta °K$) for water protons in a solution of pig plasma Benzylamine Oxidase. R_f frequency 30 MHz.

A detailed study of the R_f frequency and temperature dependence of both T_1 and T_2 for the native enzyme is being made and it should be possible to calculate a value for the number of coordinated waters as well as the distance of water from the copper in the coordination sphere. A change in the number of coordinated waters following substrate addition would indicate (though not prove) that benzylamine binds to the copper in the enzyme by displacing water molecules. N.m.r. study of the benzylamine protons when bound or unbound (as discussed in Section IIIB) should give an unambiguous answer to the question of the site of substrate binding.

The work of Mildvan and Cohn[1,35] has been referred to in the Introduction to illustrate the value of applying both e.p.r. and n.m.r. where possible. These workers have studied extensively enzymes which can be activated by Mg^{2+}, particularly kinases; the Mg^{2+} can be replaced by Mn^{2+} thus giving the system a paramagnetic probe without eliminating the catalytic activity. Two groups of kinases can readily be distinguished by magnetic resonance studies. In the first group, Mn^{2+} binds to ATP before binding to the enzyme. In the second group Mn^{2+} binds to the enzyme before binding to ATP.

Calculations of binding constants for the nucleotide and Mn^{2+} have been made as well as the distances of different protons in the nucleotide from the paramagnetic probe.

IV. Concluding Remarks

It is hoped that the great potential of magnetic resonance spectroscopic methods in the study of enzymes will now be appreciated. Both structural and dynamic information is available from pulsed n.m.r. studies; the interpretation of these spectra will be greatly facilitated when the results of other physical and chemical studies are considered simultaneously. For example, the enzyme phosphorylase b (EC 2.4.1.1)[36] has recently been studied by a combination of n.m.r., e.p.r., fluorescence and chemical reactivity methods;[37] from these studies it was concluded that transitions between four conformations of the enzyme result from activation by AMP. Observation of these conformational transitions by three independent methods encourages the belief that this conclusion is correct.

Studies on multi-enzyme complexes and on enzymes bound to membranes are in an early stage of development; "spin-labelling" methods in particular should be of great value in our understanding of these complex systems.

REFERENCES

1. Cohn, M. (1970). Magnetic resonance studies of enzyme-substrate complexes with paramagnetic probes as illustrated by creatine kinase. *Quart. Rev. Biophys.* **3**, 61-91.
2. Sheard, B. and Bradbury, E. M. (1970). N.m.r. in the study of Biopolymers. In *Progress in Biophysics and Molecular Biology*, (Butler, J. A. V. and

Noble, D. eds), **20**, pp. 187-246. Pergamon Press, Oxford.

3. Lyden-Bell, R. M. and Harris, R. K. (1969). *Nuclear Magnetic Resonance Spectroscopy*, Nelson, Belfast.

4. Farrar, T. C. and Becker, E. D. (1971). *Pulse and Fourier Transform n.m.r.* Academic Press, London and New York.

5. Raftery, M. A., Dalquiest, F. W., Parsons, S. M. and Wolcott, R. G. (1969). The use of nuclear magnetic resonance to describe relative modes of binding to lysozyme of homologous inhibitors and related substrates, *Proc. Nat. Acad. Sci. U.S.* **62**, 44-51.

6. McConnell, H. M. and McFarland, B. G. (1970). Physics and chemistry of spin labels, *Quart. Rev. Biophys.* **3**, 91-137.

6a. Smith, I. C. P. (1971). In *Biological Applications of Electron Spin Resonance*, (Bolton, J. R., Borg, D. & Swartz, H., eds), Wiley Interscience, New York.

7. Jardetzky, O. and Wade-Jardetzky, N. G. (1971). Applications of nuclear magnetic resonance spectroscopy to the study of macromolecules, *Annu. Rev. Biochem.* **40**, 605-634.

8. Dickerson, R. E. and Geis, I. (1969). *The Structure and Action of Proteins*, Harper and Row, New York, Evanston, London.

9. Stern, A., Gibbons, W. A. and Craig, L. C. (1968). A conformational analysis of gramicidin S-A by nuclear magnetic resonance, *Proc. Nat. Acad. Sci. U.S.* **61**, 734-741.

10. Gibbons, W. A., Nemethy, G., Stern, A. and Craig, L. C. (1970). An approach to conformational analysis of peptides and proteins in solution based on a combination of nuclear magnetic resonance spectroscopy and conformational energy calculations. *Proc. Nat. Acad. Sci. U.S.* **67**, 239-246.

11. Meadows, D. H., Jardetzky, O., Epand, R. M., Ruterjains, N. H. and Sheraga, A. G. (1968). Assignment of the histidine peaks in the nuclear magnetic resonance spectrum of ribonuclease. *Proc. Nat. Acad. Sci. U.S.* **60**, 766-772.

12. Wyckoff, H. W., Hardman, K. D., Allewell, N. M., Inagami, T., Johnson, L. N. and Richards, F. M. (1967). The structure of Ribonuclease-S at 3·5 Å resolution. *J. Biol. Chem.* **242**, 3984-3988.

13. Markley, J. L., Putter, I. and Jardetzky, O. (1968). High resolution nuclear magnetic resonance spectra of selectively deuterated Staphylococcal Nuclease. *Science*, **161**, 1249-1251.

14. Haestis, W. H. and Raftery, M. A. (1971). Use of Fluorine-19 Nuclear Magnetic Resonance to study conformation changes in selectively modified ribonuclease S. *Biochemistry*, **10**, 1181-1186.

15. McDonald, C. C., Phillips, W. D. and Vinogradov, S. N. (1969). Proton magnetic resonance evidence for methionine iron coordination in mammalian-type ferrocytochrome C. *Biochem. Biophys. Res. Commun.* **36**, 442-449.

16. Dickerson, R. E., Takano, T., Eisenberg, D., Kallai, O. B., Samson, L., Cooper, A. and Margoliash, E. (1971). Cytochrome C at 2·8 A resolution. *J. Biol. Chem.* **246**, 1511-1535.

17. Dayhoff, M. O. and Eck, R. V. (1967-8). *Atlas of Protein Structure and Sequence*. National Biomedical Research Foundation, Silver Springs, Maryland.

18. Allerhand, A., Cochran, D. W. and Doddrell, D. (1970). Carbon-13 Fourier transform nuclear magnetic resonance. II. Ribonuclease. *Proc. Nat. Acad. Sci. U.S.* **67**, 1093-1096.

19. Ingram, D. J. E. (1969). *Biological and Biochemical Applications of Electron Spin Resonance*. Hilger, London.
20. Berliner, L. J. and McConnell, H. M. (1971). Spin label orientation at the active site of α chymotrypsin in single crystals. *Biochem. Biophys. Res. Commun.* **43**, 651-657.
21. Monod, J., Wyman, J. and Changeux, J. P. (1965). On the nature of Allosteric Transitions: a plausible model. *J. Mol. Biol.* **12**, 88-118.
22. Koshland, D. E. and Neet, K. E. (1968). The catalytic and regulatory properties of enzymes. *Ann. Rev. Biochem.* **37**, 359-410.
23. McConnell, H. M. (1971). Spin label studies of cooperative oxygen binding to haemoglobin. *Annu. Rev. Biochem.* **40**, 227-236.
24. Moffat, J. K. (1971). Spin-labelled haemoglobins: a structural interpretation of electron paramagnetic resonance spectra based on X-ray analysis. *J. Mol. Biol.* **55**, 135-146.
25. Lanir, A. and Navon, G. (1971). Nuclear magnetic resonance studies of bovine carbonic anhydrase. Binding of sulfonamides to the zinc enzyme, *Biochemistry*, **10**, 1024-1032.
26. Taylor, P. W., Feeney, J. and Burgen, A. S. V. (1971). Mechanism of ligand binding with cobalt human carbonic anhydrase by ^{1}H and ^{19}F n.m.r. *Biochemistry*, **10**, 3866-3875.
27. Taylor, J. S., Mushak, P. and Coleman, J. E. (1970). Electron spin resonance studies of carbonic anhydrase. *Proc. Nat. Acad. Sci. U.S.* **60**, 86-91 and **67**, 1410-1416.
28. Vallee, B. L. and Williams, R. J. P. (1968). Metallo enzymes. The entatic nature of their active sites. *Proc. Nat. Acad. Sci. U.S.* **59**, 498-505.
29. Navon, G., Shulman, R. G., Wyluda, B. J. and Yamane, T. (1968). Nuclear magnetic resonance studies of the active site of carboxypeptidase A. *Proc. Nat. Acad. Sci. U.S.* **60**, 86-91.
30. Barry, C. D., North, A. C. T., Glasel, J. A., Williams, R. J. P. and Xavier, A. V. (1971). Quantitative determination of mononucleotide conformations in solution using Lanthanide ion shift and broadening n.m.r. probes. *Nature*, (*London*), **232**, 236-245.
31. Dixon, M. and Webb, E. C. (1964). Enzymes. 2nd Edition. Longmans Green, London.
32. Beinert, H. and Palmer, G. (1965). Contributions of e.p.r. spectroscopy to our knowledge of oxidative enzymes. *Advan. Enzymol.* **27**, 105-198.
33. Malmström, B. G., Reinhammer, B. and Vänngård, T. (1968). Two forms of Copper II in fungal laccase. *Biochim. Biophys. Acta*, **156**, 67-76.
34. Malkin, R., Malmström, B. G. and Vänngård, T. (1968). The requirement of the "non-blue" copper II for the activity of fungal laccase. *FEBS Lett.* **1**, 50-54.
35. Mildvan, A. S. and Cohn, M. (1970). Aspects of enzyme mechanisms studied by nuclear spin relaxation induced by paramagnetic probes. *Advan. Enzymol.* **33**, 1-69.
36. Fischer, E. H., Pocker, A. and Saari, J. C. (1970). The structure, function and control of glycogen phosphorylase. *Essays in Biochemistry*, (Campbell, P. N. and Dickens, F. eds), **6**, pp. 23-68, Academic Press, London and New York.
37. Birkett, D. J., Dwek, R. A., Radda, G. K., Richards, R. E. and Salmon, A. G. (1971). Probes for the conformational transitions of phosphorylase b. *Eur. J. Biochem.* **20**, 494-508.

The Degradation of Haem by Mammals and its Excretion as Conjugated Bilirubin

G. H. LATHE

Department of Chemical Pathology,
University of Leeds, Leeds LS2 9NL, Great Britain

I. Introduction

For a long time studies of tetrapyrrole pigments, both linear and ring compounds, have occupied an important and sometimes a central part in the development of biochemistry. That the linear tetrapyrrole pigments of bile had their origin in haemoglobin was known for more than fifty years before the structure of the oxygen receptor of haemoglobin was established by Hans Fischer's synthesis of ferrihaem in 1929. Two additional lines of biochemical interest were the nature of the extraordinary porphyrins excreted in certain diseases, and the biochemical basis of jaundice, which was placed on a more secure footing by the development of the van den Bergh reaction in 1913. Finally the rediscovery of the cytochromes, by Keilin in 1925, gave a new urgency to the study of haemoprotein structure and function at the molecular level. All of these fields have contributed to our present knowledge of the biosynthesis, biochemical functions, and degradation of haem.

Haem is the predominant tetrapyrrole pigment of mammals because of the large mass of haemoglobin and myoglobin for which it is the prosthetic group. The aim of this essay is to indicate the biochemical and cellular factors involved in the transformation of the haem of red cell haemoglobin into the pigments of bile, and their implications in human studies. As far as possible the account will follow the sequence of events as they occur, beginning with red cell destruction. This can be better appreciated after a brief description of pigment chemistry.

II. Structure and Nomenclature

A. HAEM

Porphine is a square planar aromatic ring of four pyrroles, joined by unsaturated carbon bridges. The hydrogens, β to the pyrrole nitrogens, may be substituted (the positions are numbered as in Fig. 1) to form porphyrins. The porphyrin of haemoglobin has 3 different β substituents: 4 methyl, 2 vinyl and 2 propionate. This combination is called protoporphyrin and of 15 possible arrangements that of haemoglobin has been designated protoporphyrin IX. It occurs as an iron complex with the iron offset about 0·5 Å from the plane of the macroring[1] (Fig. 2).

Metalloporphyrins of divalent metals (Mg, Zn, Cu, Fe) are formed by the displacement of two protons from 2 pyrrole nitrogens; all pyrrole nitrogens become equally bonded to the metal. Complexes with divalent metals have no net charge unless, as in protoferrohaem, there are acidic β substituents.

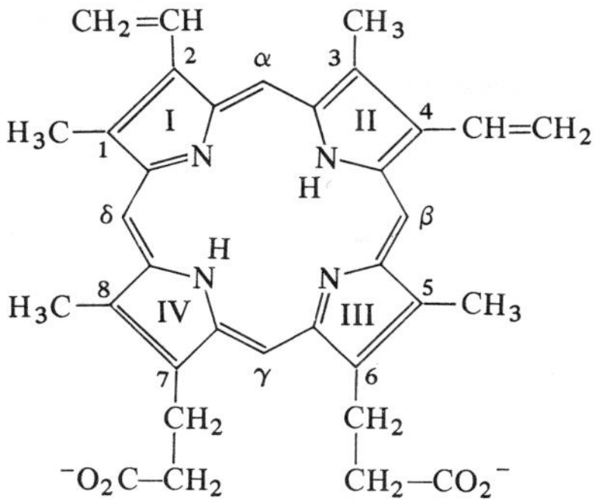

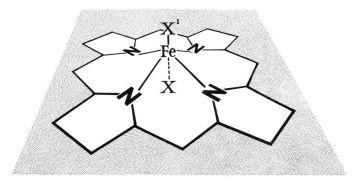

Structure showing porphyrin ring with substituents:

CH$_2$=CH (position 2, bridge α) CH$_3$ (position 3)

H$_3$C— (ring I, position 1) N

II, position 4 —CH=CH$_2$

N H

δ

H N

β

H$_3$C— (position 8, ring IV) N N III —CH$_3$ (position 5)

H

γ

position 7 CH$_2$ CH$_2$ (position 6)

$^-$O$_2$C—CH$_2$ CH$_2$—CO$_2$$^-$

Fig. 1. Protoporphyrin, with conventionally named bridges and numbered positions β to the pyrrole nitrogens.

Fig. 2. Haemochrome showing the two ligands, X and X^1, and iron offset from the plane of the porphyrin.

Iron, having 6 co-ordination positions, can accept one or two more lone pairs of electrons to form a haemochrome (syn. haemochromogen) (Fig. 2). In the cytochromes the iron porphyrins are haemochromes, sharing 2 pairs of electrons with imidazole residues of the protein. Ferrohaemochromes have more distinctive, 3 band, spectral characteristics than ferri-complexes, or uncoordinated compounds, and for this reason estimations, or identifications, are often made by conversion to ferrohaemochromes (usually of pyridine). In haemoglobin and myoglobin only one of the coordination positions of iron is occupied by an imidazole residue and various weak field ligands (H_2O, OH^-, O_2) compete for the 6th position. These can readily be displaced by strong field ligands (CO, NO). Fe^{3+} may also form a metalloporphyrin complex but the loss of an electron from ferrohaem produces two changes. Firstly, the electronic structure of the porphyrin is altered (spectral changes occur; ferrihaemoglobin does not co-ordinate with oxygen), and secondly, the ferrichelate has a net charge which must be balanced in crystals or solution by an anionic ligand. Ferrihaemochrome occurs naturally as the prosthetic group of some hydroperoxidases, e.g. catalase. The electron transport function of the cytochromes, however, requires inter-conversion of the ferri- and ferrohaemochromes.

Ferroprotoporphyrin IX is usually referred to as haem, and ferrihaem as haemin or haematin. The older terms "acid haematin", for compounds such as chloroferrihaem (syn. ferrihaemchloride), and "alkaline haematin" for hydroxy-ferrihaem, are sometimes still used and in some papers "haemin" refers only to the former and "haematin" to the latter. It is an old custom to use the prefix met- (in the sense of change) to indicate oxidation, as methaemoglobin, methaemalbumin. It will be clear why the Enzyme Commission of the I.U.B. recommended that "when the state of oxidation of the iron atom is to be specified ferrohaem and ferrihaem (and ferrylhaem) should be used."[2] That convention will be followed here. The unwary should note, however, that in the literature haem is often used in two senses. Firstly, there is a general sense in which the oxidation state is immaterial and both forms are included (e.g. the haem compounds of the respiratory chain). Secondly, the oxidation state may be important but not yet known (e.g. Tenhunen et al.[3] called the enzyme which acts on ferrihaem in a reducing medium, haem oxygenase).

Some haem proteins have substituents in positions β to the pyrrole nitrogens, which differ from those of protohaem, e.g. in cytochrome a the methyl side chain in the 7 position has been oxidized to a formyl residue. Haem a is used in this sense.[2] Nothing is yet known of the pathways of degradation of haem a and of other cytochromes.

B. BILE PIGMENTS

Chemical or biochemical rupture of the macroring of haem may take place at any of the 4 methine (syn. methyne, methene, methenyl) bridges. These have been designated systematically beginning with α opposite the propionic acid side chains (Fig. 1). The order of the substituents in the resulting linear tetrapyrrole pigments will depend on which methine bridge in the parent porphyrin has been ruptured. The sequence of β-substituents is indicated by reference to the deleted methine bridge—thus breaking photoporphyrin IX at the α-methine bridge produces biliverdin IXα (Fig. 4).

In bilirubin (Fig. 4), which is a biladiene, the central methylene (syn. methyl) bridge interrupts the conjugated system extending through rings A, B, C and D in the parent bilatriene, biliverdin. The methylene bridge of bilirubin is less stable than the methine bridge of biliverdin: it is the site of rupture in the diazo reaction (p. 122) and of fragmentation in mass spectrometry.[4] Rupture of the central bridge produces dipyrrole pigments which vary with the parent bilirubin, and may be used to identify this.[5]

III. Haem-Apoprotein Interaction

Dissociable haem compounds might be expected to be in equilibrium *in vivo*, with a pool of "free" ferri- and ferrohaem: ferrohaem being autoxidizable cannot exist in plasma, and perhaps not in cells. Evidence for the dissociation of haem from apoproteins is drawn from the exchange of $[^{59}Fe]$ haem between haemoproteins.[6] Haemoproteins can be arranged in order of the speed with which haem exchange occurs or their affinity for haem, beginning with the least dissociated:

haptoglobin-haemoglobin	(ferrohaem)
oxyhaemoglobin	(ferrohaem)
haemoglobin A	(ferrohaem)
haemoglobin F	(ferrohaem)
methaemoglobin	(ferrihaem)
haemopexin	(ferrihaem)
methaemalbumin	(ferrihaem)

Thus haemoglobin and plasma albumin undergo little haem exchange but methaemoglobin and plasma albumin exchange rapidly and approach equilibrium in 3 hours.

IV. Sites of Haemoglobin Breakdown

Administration of haemoglobin or of altered red blood cells leads to an increase in bile pigment secretion. The transformation to a yellow pigment can be seen in tissues into which blood had escaped. Cells which can take up haemoglobin and form bile pigment occur in almost all tissues but this cell type is particularly prevalent in the spleen, bone marrow, liver, and lymph glands. Some of these cells, variously called macrophages, histiocytes, reticular cells and, in the liver, Küpfer cells, are capable of enlargement, migration and phagocytosis. Aschoff referred to them as the reticulo-endothelial system and today they are often considered to form a distinct, but widely dispersed, organ. In mammals, hepatectomy is followed by an accumulation of bile pigment in plasma indicating that the liver is necessary for secretion of bile pigments but not for their formation.

The site of red cell breakdown and haemoglobin catabolism in intact animals, and under normal conditions, is more difficult to establish. Several organs and several mechanisms may be involved. By administering isotope-labelled haemoglobin precursors, such as glycine,[7] or more recently δ-aminolaevulinate,[7a] and measuring the appearance of the label in the haem of haemoglobin and later in the faecal bile pigments (Fig. 3), it was shown that human erythrocytes have a life span of about 100 days. In the rat, red blood cells are broken down more randomly.

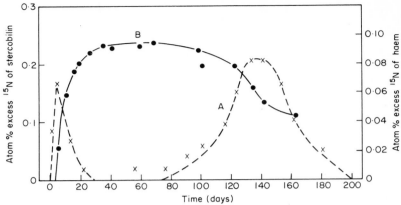

Fig. 3. The relative amount of [15N] in faecal stercobilin (curve A) and in haem of haemoglobin (curve B) following administration of [[15N]glycine to a normal man. A small "early-labelled peak" of faecal pigment and a larger one at the end of the erythrocyte life are shown. Taken from Gray et al. [7]

A. TRANSPORT PROTEINS—HAPTOGLOBINS

Normal plasma contains about 4 μmol/l haemoglobin or less ($<$ 10 mg/100 ml), but most of this may be an artifact of blood sampling. There are one or more, genetically determined, haemoglobin-binding proteins in plasma. Type 1–1 haptoglobin has a molecular weight of about 85,000 and binds one mole of haemoglobin. Plasma contains about 12 μmol/l of haptoglobin (equivalent to about 75 mg/100 ml haemoglobin). The complex (but not haemoglobin alone) is larger than the pores in the renal glomerular membrane which may explain why haemoglobinuria does not occur unless the plasma haemoglobin concentration exceeds the binding capacity of plasma haptoglobin. Inherited deficiency of haptoglobin is common in parts of Africa (occurrence as high as 30%) and appears to confer no disadvantage.

Rabbit plasma has very little haptoglobin and the rabbit is therefore useful for comparing the disposal of haemoglobin and of the haemoglobin-haptoglobin complex. Murray et al. [8] found that intravenously-administered haemoglobin was removed rapidly ($t\frac{1}{2}$ of 25 min) and that it passed about equally to kidney and liver. In contrast, haptoglobin-haemoglobin ($t\frac{1}{2}$ of 100 min) went mainly into the liver. Presumably the complex is taken up intact for the $t\frac{1}{2}$ of haptoglobin alone is 40 times that of the complex. Because of this rapid removal of the complex and slow replacement of haptoglobin, plasma haptoglobin is low in diseases in which intravascular haemolysis occurs.

Transfused red blood cells are removed rapidly from the circulation if the cells have been "sensitized" by attaching antibodies. Ostrow, Jandl and Schmid[9] noted that these cells are taken up about equally by spleen and liver in the rat. In contrast, intravenously administered haemoglobin can be traced mainly to liver and kidney. If very large amounts of haemoglobin are given it appears in the urine and much accumulates in the kidney, where it is probably broken down in the proximal tubule after glomerular filtration and reabsorption.

In rats sensitized red blood cells and labelled haemoglobin are removed from the circulation at approximately the same rate, 70% to 95% being removed in 2 h.[9] Following administration of labelled haemoglobin or erythrocytes, bilirubin begins to appear in the bile after about 30 min, the peak rate of excretion is at about 2 h and 50% has been excreted by 3 h. This is slower than the biliary excretion of intravenously injected bilirubin (0.4 μmol) which begins almost immediately; half is recovered in 20 min and all of it in 2 h. The net recovery after small amounts of haemoglobin is 100%, as with injected bilirubin. With large amounts of haemoglobin (exceeding the binding capacity of plasma haptoglobin), or sensitized red cells, only about 75% is recovered as bilirubin

B. TRANSPORT PROTEINS—METHAEMALBUMIN AND HAEMOPEXIN

If there is sufficient intravascular haemolysis for the plasma haemoglobin to exceed the binding capacity of haptoglobin, methaemalbumin, a complex of ferrihaem and plasma albumin, appears in the plasma. This does not happen on adding haemoglobin to plasma *in vitro*; some additional factor is required. Albumin binds up to 2 moles of ferrihaem but it is not known whether the two sites are identical.

In 1958 Neale, Aber and Northam[10] noted that if sera from patients with intravascular haemolysis were examined by paper electrophoresis, peroxidase activity was found in the β region. This led to the discovery of a specific haem-binding protein, haemopexin, in addition to albumin.

Haemopexin is a β_1-glycoprotein of 70,000 molecular weight (man). It has one binding site for haem[11] with a much higher affinity than has albumin.[12] On reduction with dithionite the complex gives the 3-band absorption which is characteristic of ferrohaemochrome.[13] Haemalbumin can be reduced but not to a ferrohaemochrome. Both reduced complexes are autoxidizable.

Recent work by Drabkin[13] suggests a biological role for haemopexin. In the isolated perfused rat liver excellent bile pigment production from ferrihaem occurred when a haemopexin concentrate was added to the perfusate, but not otherwise. Thus haemopexin may be necessary for haem to be taken up and degraded by the liver.

C. SPECIALIZATION OF LIVER CELLS

Which cells in the liver are responsible for haemoglobin breakdown and haem degradation? Hepatocytes account for only about 65% of the liver cells and the reticuloendothelial cells lining liver capillaries (i.e. sinusoidal cells, Küpfer cells; Fig. 8) account for an important fraction of the remainder. It has been shown by electron microscopy and histochemical staining that following intravenous administration of haemoglobin to rats both hepatocytes and Küpfer cells take up haemoglobin by pinocytosis.[14] Within 30 min the vacuoles containing haemoglobin can be stained for lysosomal enzymes (acid phosphatase, β-glucuronidase and glucosaminidase) and the number of lysosomes falls.[15] Bissell, Hammaker and Schmid[16] used changes in an enzyme, haem oxygenase, (p. 118) to determine where haemoglobin and red cells were taken up. They separated the reticuloendothelial cells from the hepatocytes of rat liver after intravenous injection of sensitized erythrocytes or haemoglobin. After giving red cells, induction of haem oxygenase occurred only in reticuloendothelial cells. Following haemoglobin the main response was in the hepatocytes.

D. SUMMARY

The role of haptoglobin may be to direct plasma haemoglobin towards the hepatocytes rather than to the kidney. The biological advantage of this is not one of reduced wastage, as haem is discarded and iron is retained in either case. It may be a means of directing haemoglobin to the main haem degrading system, for haem is recovered quantitatively as bilirubin from the liver. Sensitized red blood cells are taken up by the reticuloendothelial cells of the liver and spleen and an important part of the haem is diverted from the bilirubin pathway. The delay, due to degradation and reduction, before the haem of haemoglobin or of red cells is excreted as bilirubin is 30 min or more, while the lag in the secretion of injected bilirubin is measured in seconds.

The normal disposal of old red blood cells is more difficult to define. Much indirect evidence suggests a major role for the spleen. Splenectomy has been shown to induce haem oxygenase of liver.[17] The occurrence of bilirubin in plasma suggests a flow of pigment to the liver. A similar flow of haem has not been demonstrated. Perhaps haptoglobin and haemopexin have been evolved to meet abnormal conditions of haemolysis and play no part in the normal transfer of tetrapyrrole pigments.

V. Intracellular Haem

Much attention has been given to the mechanisms of protoporphyrin generation, its control and the incorporation of iron. Most cells require haem for respiratory cytochromes, and in the liver relatively large amounts are needed for synthesis of other haemoproteins, especially microsomal cytochrome P-450 which probably has a half-life of less than 12 h in the rat.[18]

Liver catalase (EC 1.11.1.6) appears to be concentrated in peroxisomes but the sap contains another haemoprotein, tryptophan oxygenase (EC 1.13.1.12).

The apoenzyme of tryptophan oxygenase must be present in the liver cell sap for addition of ferrihaem *in vitro* increases the enzyme activity.

Granick and Drabkin have suggested, independently, that haem is part of an inhibitory "loop" controlling the biosynthesis of the rate-limiting enzyme of protoporphyrinogen production, δ-aminolaevulinate synthetase. Both the synthetase and the haem-producing enzyme, haem ferrochelatase (EC 4.99.1.1), are localized in the mitochondria. Since haemoproteins are assembled on cytoplasmic ribosomes, ferro- or ferrihaem must diffuse, or be transported, through the mitochondrial membrane and the sap to the endoplasmic reticulum. More precise knowledge of the amount of free haem is greatly needed.

VI. Mechanism of Haem Degradation

A. LEMBERG

In 1930 Warburg and Negelein made "green haemin" from ferrihaem by oxidation in the presence of reducing agents. Fischer considered that the change in colour was produced by interruption of the conjugation between rings I and IV of ferrihaem. He suggested that a methine bridge was oxidized forming an oxoporphyrin (cf. Fig. 5). Lemberg[19,20] oxidized pyridine haemochrome (reduced with ascorbic acid or hydrazine) to a crystalline green pigment "verdohaemochrome" (λ_{max} 640-660 nm), from which biliverdin could be obtained by treatment with acid. His group showed that green pigments could be formed from haemoglobin by combined oxidation (H_2O_2) and reduction (ascorbic acid). The product was called green or verdo-haemoglobin, and also choleglobin; the absorption (λ_{max} 610-630 nm) differed from verdohaemochrome. Lemberg considered that in these compounds the α-methine bridge of protoporphyrin was replaced by an oxygen bridge, iron being retained. This extremely confusing period of bile pigment chemistry was superseded by more modern biochemical and chemical techniques in the late 1950's. There were two main approaches. Firstly, a renewed search was made for tissue extracts which would produce biliverdin from haemoglobin, or a haemoglobin derivative. Secondly, synthetic porphyrin congeners were tested, *in vivo* and *in vitro*, to see whether they could be intermediates in bile pigment formation from haemoglobin (p. 119).

B. NAKAJIMA

In 1958 Nakajima reported that crude liver homogenate would convert pyridine ferrohaemochrome to a green product (λ_{max} 656 nm). The reaction appeared to depend on an enzyme in the sap and a low molecular weight reductant, but there were a number of unexpected features.[21] The reaction occurred mainly in liver and kidney, but not in spleen and bone marrow. Of the possible natural substrates the complex of oxyhaemoglobin with haptoglobin appeared to be the most reactive.

The supposed difference between verdohaemochrome and the product obtained by Nakajima was soon disproved,[22,23,24] and it was shown that liver ascorbic acid accounted for most, but not all, of the low molecular weight reductant.

A haem-degrading system must meet the important criterion that the sequence of side chains (β-substituents on the pyrrole rings) be IXα (Fig. 4) as in natural biliverdin and bilirubin. Although Nakajima and Gray[25] concluded from permanganate degradation of the Nakajima product that it was largely IXα, Nichol and Morell,[5] using the more accurate technique of mass spectrometry to identify the products formed by fragmentation at the central methine bridge, found that rupture at β or γ bridges also occurred with pyridine haemochrome and hydrazine or ascorbic acid, and with the Nakajima system (guinea-pig liver homogenate using pyridine haemochrome).

C. Ó CARRA AND COLLERAN

Ó Carra and Colleran[26] examined the sequence of side chains by a thin layer chromatographic system which separates the four isomers of the dimethyl esters of biliverdin IX, making quantitative estimation possible. Extracts of liver acting on pyridine haemochrome yielded only 30-33% of the α-isomer, as also did coupled oxidation by O_2 and ascorbate. Colleran and Ó Carra[24] confirmed these findings with Rüdiger's[27] highly sensitive technique of degradation with chromate. They noted that treatment of ferrihaem (which does not undergo coupled oxidation with reductants under physiological conditions) with tissue extracts usually gave the products of random bridge rupture. However, some haemoproteins, for example, myoglobin, formed largely biliverdin IXα. If the myoglobin was denatured with urea, random degradation occurred. Ó Carra and Colleran[26] related this to the haem binding site of the protein and to the hydrophobic region in which the α-methine is located. Since only the α-bridge is inaccessible they concluded that "the specific cleavage of the α-methine bridge must therefore be a positive effect of the haem-binding site rather than a masking of the other three bridges". Human haemoglobin gave a proportion (40%) of β-fragmentation probably due to a more non-polar environment for the β-bridge than in myoglobin, while catalase behaved similarly; cytochrome P-450 resembled myoglobin. A later study[28] showed that complexing human haemoglobin with haptoglobin considerably reduced the proportion of β-methine oxidation.

A further surprising observation was made by Ó Carra and Colleran.[26] Ferrihaem, which is relatively stable alone is broken down more than 10 times faster in the presence of myoglobin and the latter confers α-specificity on the products. They suggested that an intracellular system of haemoproteins which exchange haem groups (possibly including haemopexin and cytochrome P-450) may serve to provide the specificity of enzymes, the haem-binding sites being equivalent to active sites. Such a system might be expected to yield some biliverdin IXβ with human haemoglobin. They pointed out[29,30] that biliverdin reductase acts so slowly on this isomer that it might be diverted into a non-bilirubin channel.

D. TENHUNEN, MARVER AND SCHMID

The unusual (unbiological) characteristics of the Nakajima system led Tenhunen et al.[3] to re-examine tissue extracts. They found, in microsomal fractions of spleen, liver, brain, kidney and lung, a NADPH- and O_2-requiring system for the conversion of ferrihaem to biliverdin (Fig. 4). (In practice biliverdin was usually measured as bilirubin following the addition of excess biliverdin reductase and NADPH; see p. 119). Five minutes at 56°C destroyed

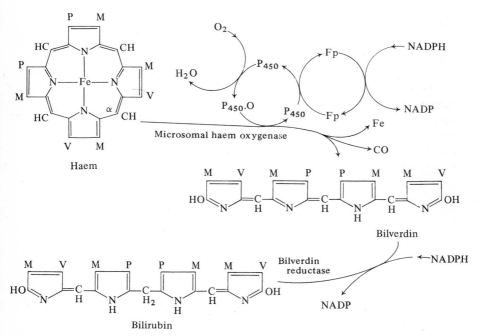

Fig. 4. Degradation of haem to biliverdin by haem oxygenase and reduction to bilirubin according to Tenhunen et al.[3] Biliverdin and bilirubin have been drawn in the bis-hydroxypyrrole form. Taken from Schmid.[130]

the microsomal activity. Methaemoglobin and the haem complexes of α- and β-chains of haemoglobin were degraded but not oxyhaemoglobin or myoglobin. Iron-free porphyrins were not attacked. They regarded this system as a mixed-function oxygenase, since it required both NADPH and O_2, and called it haem oxygenase, using haem in the general sense. Equal amounts of bilirubin and CO (from the α-methine bridge) were produced. Inhibition by CO suggested participation by cytochrome P-450. It was further shown that photoactivation of the system released the CO inhibition, and the spectral characteristics of cytochrome P-450 and the photoactivation spectrum were very similar.[31] It differed from the photoactivation spectrum of a CO-ferrihaem system (tryptophan oxygenase). On the other hand, a number of substrates of drug-metabolizing enzymes for which cytochrome P-450 is the terminal stage (hexobarbital, aminopyrine) and SKF525A, which blocks many drug-metabolizing enzymes, did not inhibit haem oxygenase. This may be due to specialization of the cytochrome P-450 enzymes of microsomes and may also explain why a 6-10 fold increase in liver oxygenase is not associated with an increase in the total amount of P-450.[17] The stoicheiometry of O_2 consumption and biliverdin production was said to be consistent with a mixed-function oxidation but the acccuracy was not claimed to be very high.

The mechanism of the Tenhunen reaction has been partly elaborated. The bilirubin was mainly IXα (by mass spectrometry and permanganate degradation). $^{18}O_2$ and $H_2{}^{18}O$ were used to define the source of the oxygen. CO and the 2 α-oxygen atoms in bilirubin were labelled from O_2 (Fig. 4) but not from H_2O.[31] The preferred substrate appeared to be ferrihaem, and protohaem was much more active than deutero- or coprohaem. Protoporphyrin itself was inactive. Those haemoproteins from which the haem group readily dissociates (e.g. methaemalbumin, methaemoglobin) were active while those in which haem is more tightly bound (oxyhaemoglobin, myoglobin, haptoglobin-haemoglobin complex) were almost inactive. Haem bound to haemopexin was degraded.

One of the striking characteristics of the system was the extent to which it could be induced.[17] Thus, following administration of haemoglobin or methaemalbumin the activity of spleen increased (2-7 fold) and also in haemolytic anaemia (3-5 fold), while the liver responded to splenectomy (2-3 fold) and zymosan administration (6-10 fold). Induction also occurred in proximal renal tubules,[32] peritoneal macrophages,[33] lung[17] and two types of liver cell[16] (p. 114). Fasting and administration of hormones mediated by cyclic AMP increased the activity of liver enzyme.[34]

It is not possible at present to be entirely certain whether or not the reaction is enzymic. The usual criterion of heat lability does not distinguish between the concepts of Ó Carra and Tenhunen. Ó Carra suggests that cytochrome P-450 is not acting as a hydroxylase, but rather that the apo-protein is activating the α-methine of haem. Haem IXα is the coenzyme of cytochrome P-450 and therefore the α-methine activation or fission must take place, according to Ó Carra, *in situ*. As all cytochrome P-450 would be expected to be equally effective according to this theory, provided that it was equally dissociated, it seems surprising that the substrates for drug-metabolizing enzymes do not diminish ring fission. The absence of this effect suggests a degree of cytochrome P-450 specificity. The experiments with CO inhibition would appear compatible with either theory. On the other hand the Tenhunen interpretation is strongly supported by the necessity of NADPH for conversion of haem to biliverdin.

E. *IN VIVO* STUDIES OF POSSIBLE INTERMEDIATES– KONDO, NICHOLSON, JACKSON AND KENNER

Although it is not possible to define the intermediates by administering them it should be possible to exclude as intermediates those substances which do not yield bilirubin. Kondo et al.[35] tested tritiated synthetic oxyporphyrins and their iron chelates (Fig. 5). Only meso compounds were available (Tenhunen et al.[3] found that mesoferrihaem slightly inhibited haem oxygenase.) In 24 h, 30% of injected α-oxomesoferrihaem and 4% of the β-oxo isomer were excreted in the bile mainly as bilirubin or mesobilirubin; α-oxoporphyrin was rapidly excreted in the bile, but probably in an unaltered form; β-oxoporphyrin was excreted very slowly. This would suggest that the presence of iron in the structure is essential for normal degradation and that the α-oxoferrihaem is much more rapidly degraded and excreted than the β-oxoferrihaem. In control experiments it was shown that the rate-limiting factor was neither the reduction of biliverdin IXβ nor its excretion, both of which proceeded much more rapidly than after administration of β-oxo-mesoporphyrin. As a possible explanation of these findings Kondo et al. pointed out that both α- and β-oxomesoferrihaems readily isomerize to the corresponding hydroxymethine form and are oxidized to bile pigments *in vitro* by atmospheric oxygen (Jackson et al.[36]). They suggest that a mole of oxygen is introduced non-enzymically across the α-oxomethylene bridge, for which there are chemical models. This would satisfy the $^{18}O_2$ labelling experiments of Tenhunen et al.[31] This mechanism and the alternative paths of a mixed-function oxygenase are shown in Fig. 5.

VII. Biliverdin Reductase

In 1936 Lemberg and Wyndham observed that liver mince, to which biliverdin had been added, turned yellow. Singleton and Laster[37] and Tenhunen et al.[38] (Fig. 4) have studied this biliverdin-reducing system, but disagreed on the coenzyme requirement. Colleran and Ó Carra[29,30] attributed this to the use of different concentrations of coenzyme and biliverdin. At 27 μmol/l biliverdin, the K_m for NADPH was about 5 μmol/l compared with more than 500 μmol/l for NADH. The K_m for biliverdin was 1-2 μmol/l with NADH and below 0·2

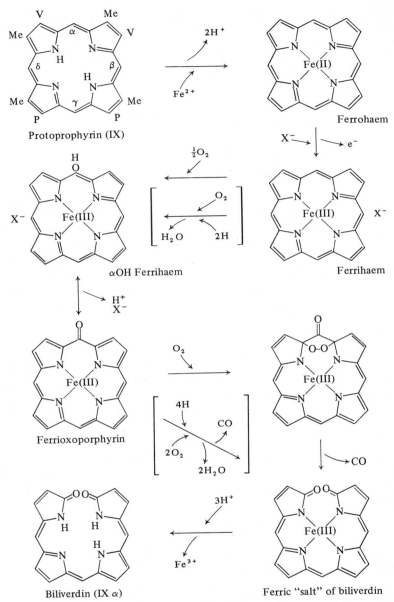

Fig. 5. The formation of ferrohaem, and its degradation to biliverdin according to Kondo *et al.*[35] Transitions which might be catalysed by the mixed function oxygenase of Tenhunen *et al.*[3] are shown in parentheses. The figure has been modified from Kondo *et al.*[35] Pyrrole substituents Me, V and P are omitted for clarity.

μmol/l with NADPH. They found a high specificity for biliverdin IXα, and argued that if the haem ring was opened at the β, γ or δ bridges the products would probably not appear as bilirubin.

Tetrapyrrole pigments must move from one cell compartment to another. Thus, biliverdin is formed in the endoplasmic reticulum, reduced to bilirubin in the sap and finally (p. 130) bilirubin is conjugated in the endoplasmic reticulum.

VIII. Bilirubin

A. PROPERTIES AND STRUCTURE

The sodium and potassium salts of this dicarboxylic acid are soluble but at a neutral or acid pH bilirubin is remarkably insoluble. Overbeek et al.[39] titrated bilirubin and found a single pK_a at 8·0 (cf. pK_a of propionic acid 4·9). They estimated that at pH 7 and 8 the solubilities were 2 and 200 μmol/l respectively (0·1 and 11·7 mg/100 ml). The aqueous insolubility is remarkable for a compound that has so many potential sites for hydrogen bonding. Fog and Jellum[40] suggested that this might be explained by internal hydrogen bonding. They considered two structures, one of which (the less strained) is shown in structure A, Fig. 6. They compared the infrared spectrum of mesobilirubin with that of its dimethyl ester in which hydrogen bonding through the carboxyls would be blocked. Hutchinson et al.[41] concluded from the infra-red and NMR spectra of bilirubin that there was strong internal hydrogen bonding. They favoured structure B, Fig. 6 because it should not lower the carbonyl stretching vibration.

Bilirubin has other unusual properties—it is not strongly basic, and forms α-hydroxy derivatives and divalent metal complexes with difficulty. Many linear tetra- and dipyrrole- compounds like biliverdin are considered to be partly in the lactim (bis-pyrrolenine, pyrrolidene, or hydroxypyrrole) form, as shown in Fig. 4. The lactim structure would be strongly basic, allowing for formation of hydrochlorides because the lone pairs of electrons on the pyrrolenine nitrogen can be shared with protons. However, the lactam structure with no pyrrolenine nitrogens seems more appropriate for bilirubin (Fig. 6) as it is weakly acidic.[39]

The lactim structure of biliverdin also allows for complex formation with divalent metals, which requires that one or two protons are displaced from lactam ring nitrogens and that 1 or 2 additional lone pairs of electrons are available on pyrrolenine nitrogens for co-ordination. This would be possible in both the lactam and lactim structures of biliverdin but not in the lactam (bis-pyrrolenone) structure of bilirubin (Fig. 6). Bilirubin does not form divalent cation complexes readily, though Ó Carra[42] obtained a non-fluorescent, acid-reversible, zinc complex, and Hutchinson et al.[43] isolated a zinc bilirubin complex from dipolar aprotic solvents.

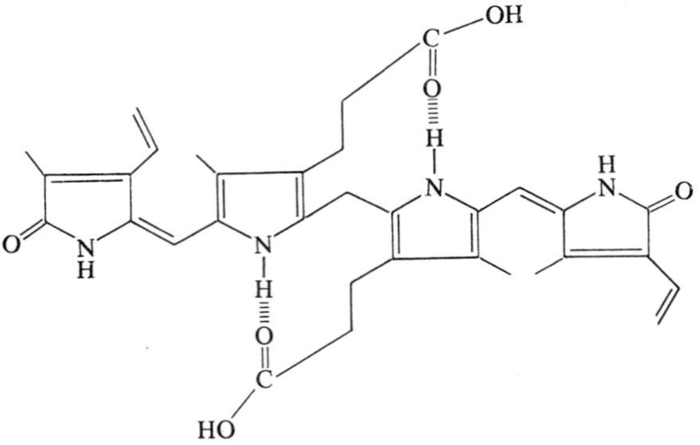

Fig. 6. Stabilization of bilirubin by intramolecular hydrogen bonding.

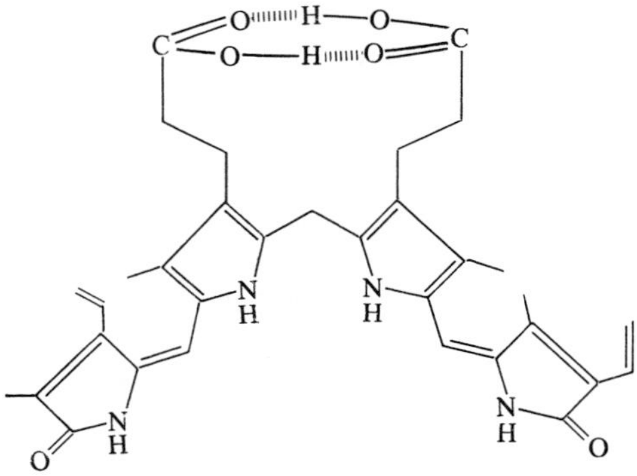

The predominance of a bis-pyrrolenone structure in bilirubin would also explain the difficulty in forming α-hydroxy derivatives. However, isomerization of bilirubin to a bis-hydroxypyrrole form is possible as methoxy derivatives have been prepared.[44,45] Hutchinson et al.[41] found that in deuterated dimethyl-sulphoxide bilirubin exhibited the n.m.r. signal at 5·25 ppm, which had been attributed[44] to enolic protons of the lactim tautomer. Thus the unusual properties of bilirubin are best explained by internal hydrogen bonding of a bis-pyrrolenone structure.

B. DETERMINATION OF BILIRUBIN

Van den Bergh (1913) applied Ehrlich's diazo reaction to the quantitative estimation of bilirubin in serum. Overbeek et al.[46] showed that coupling of bilirubin with diazotized sulphanilic acid occurred in 2 steps (Fig. 7).

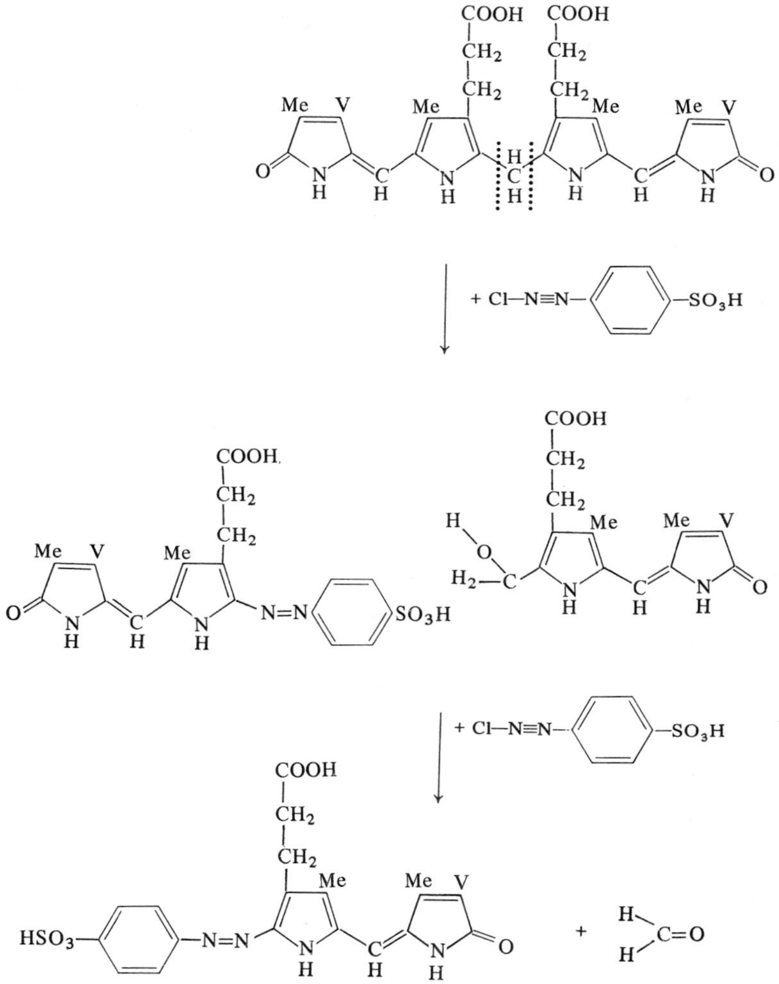

Fig. 7. The determination of bilirubin by coupling with diazotized sulphanilic acid.

Hutchinson *et al.*[47] have identified as formaldehyde the central methylene carbon, which is displaced in the reaction. Because the methyl and vinyl substituents of bilirubin are asymmetrically placed there are 2 isomeric azopigments. These behave identically in partition chromatography but separate on thin layer.[48,49]

Van den Bergh discovered that the pigment of bile reacted "directly" in aqueous solution whereas certain sera, and reagent bilirubin, reacted very slowly unless an accelerator, like ethanol, was added. Determination of the relative amounts of "direct" and "indirect" bilirubin are usually made at a low pH to minimize the coupling of bilirubin. The "indirect" reaction of bilirubin may be due to a less reactive δ-methylene bridge due to the internal hydrogen bonding of bilirubin.[40,41] The methyl and glucuronic acid esters of bilirubin react directly, possibly because the internal hydrogen bonding is interrupted.

Standard methods for determination of bilirubin in clinical biochemistry laboratories are now available.[50] For study of the azopigments of direct-reacting (conjugated) bilirubin, diazotized p-iodoaniline[51] or diazotized ethyl anthranilate[52] can be used, followed by separation of the azopigments on thin layer.

C. PLASMA ALBUMIN–BILIRUBIN COMPLEX

In man the plasma contains 0·2-0·8 mg/100 ml (3-14 μmol/l). The concentration in the rat is so low that it cannot be estimated accurately. Although the usual diazo determinations suggest that human plasma contains some direct-reacting bilirubin, a critical study of azopigments from normal human sera, showed conjugated pigment in only 1 of 36 sera.[51]

Although more than 30 molecules of bilirubin can combine with one of plasma albumin if pigment is added to sera, this is of no biological interest as the pigment:albumin ratio in man very rarely rises above 1:1.

Broderson and Bartels[53] developed a very sensitive method of bilirubin determination based on oxidation to biliverdin with peroxidase and H_2O_2. Since albumin-bound bilirubin does not react, the unbound bilirubin of plasma can be determined. The dissociation constant for the first bilirubin-binding site of human albumin is 7×10^{-9} (ref. 54). The second site is different, for its dissociation constant is about 300 times greater. Bovine and rat albumins have lower affinities than human albumin.

D. UPTAKE OF BILIRUBIN BY THE LIVER

The liver circulation of a normal man is about 1200 ml blood per minute. As only 40 ml of plasma must be cleared of bilirubin per minute the percentage removal is low. The half-life of a small intravenous dose of radio-bilirubin is 17·7 min[55] in man and about 5 min in the guinea pig.

Unlike most cells of the body the hepatocytes are bathed in fluid of approximately the same concentration as plasma (Fig. 8). The concentration of unbound bilirubin in plasma is several hundred times too low to account for the uptake, and it is doubtful whether pinocytosis could account for 40 ml of plasma per min. As normally only 1 in 60 plasma albumin molecules has attached bilirubin it is possible the carrier molecules are selectively taken up—perhaps as a result of a conformational change. A rate of uptake of about 30 mg albumin per min would be required. Another possibility is that unbound bilirubin is transported into the parenchymal cells, so that the albumin-bilirubin complex can dissociate by about 10-20% during circulation through the liver. A number of attempts have been made to delineate an uptake mechanism, usually by the use of inhibitors.[56,57,59]

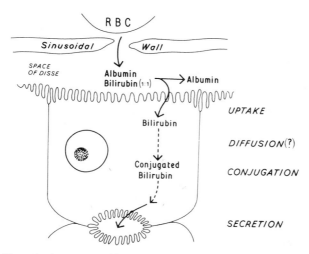

Fig. 8. Stages in the passage of bilirubin from the blood to the bile canaliculus.

E. LIGANDIN

Attention has recently turned to the possibility that the concentration gradient from plasma to liver is produced by a liver protein which has higher affinity for bilirubin than has plasma albumin. Levi *et al.*[60] fractionated the liver sap (or homogenate), to which bilirubin (or a cholephilic dye such as bromosulphthalein) had been added, by passage over Sephadex G75 (Fig. 9). Most of the protein (fraction X) was distinct from 2 small protein bands (Y and Z) which contained the dye and pigment. By administering [³H] bilirubin *in vivo* the "Y" protein was shown to have greater affinity for bilirubin than "Z". "Y" protein had already attracted the attention of two other groups interested in the binding of carcinogenic aminoazo dyes, and corticosteroid metabolites. The three groups of workers[61] have now suggested that the protein which binds organic anions be called "ligandin". It is a basic protein of 44,000 molecular weight, consisting of two subunits.[62] Levine *et al.*[63] have shown that ligandin occurs only in land animals and in amphibia it first appears during metamorphosis. Thus there is little in the liver of newborn *Macaca mulatta*[64] and it develops during the first 3 weeks of life.

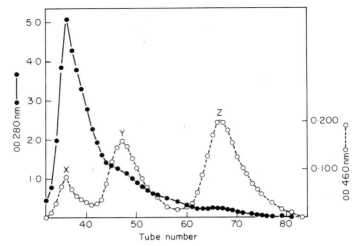

Fig. 9. Fractionation of the liver cell sap of a Gunn rat, on Sephadex G 75. The protein (●——●) and pigment (○——○) in the fractions are shown. Because the Gunn rat is not able to clear bilirubin from the liver it was not necessary to add exogenous bilirubin since the two proteins (Y and Z) with affinity for organic anions were already labelled. Taken from Levi et al.[60]

IX. Conjugated Bilirubin

A. BILIRUBIN DIGLUCURONIDE

The azobilirubin pigments (cf. Fig. 7) are used both in the determination of bilirubin and in defining the nature of direct-reacting bilirubin. In 1953, the direct- and indirect-reacting bile pigments were separated by reverse-phase chromatography (Fig. 10); Billing found that direct reacting pigments gave two types of azopigment (Fig. 10). In 1957 azopigment B was shown to be an alkali-labile,[65] β-glucuronidase hydrolysable[65,66] derivative of azobilirubin A. Three types of linkage between bilirubin and glucuronic acid are possible—an ester through the carboxyethyl side chains, an ether in the α position of rings A and D, or an N-glucuronide. An ester link was suggested by the alkali-lability and its hydrolysis by hydroxylamine.[67] Formation of N-glucuronides[67] can be excluded by their marked acid-lability. The fragmentation pattern in mass spectrometry has now shown that glucuronic acid in azopigment B is attached to the carboxyl[68] and Jansen and Billing[69] have confirmed this with the amide formed from azopigment B by ammonolysis. Since only azopigment B was obtained from the more polar direct-reacting bile pigment, pigment II (Fig. 10), it was considered to be a diglucuronide.[65,66] Talefant[70] reached a similar conclusion at the same time.

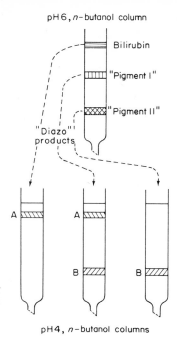

pH 6, *n*-butanol column

Bilirubin

"Pigment I"

"Pigment II"

"Diazo"
products

A

A

B

B

pH 4, *n*-butanol columns

Fig. 10. The fractionation of serum pigments, from a patient with obstructive jaundice, on a reverse-phase chromatography column. The pigment bands were eluted, coupled with diazotized aniline and rechromatographed, giving the two main types of azo pigment.[65a]

B. PIGMENT I AND BILIRUBIN MONOGLUCURONIDE

Extracts of icteric sera contain a pigment, pigment I, of polarity intermediate to those of bilirubin and pigment II[65,65a] (Fig. 10). Pigment I has the appropriate glucuronic acid content for the monoglucuronide but rechromatography[65,71] yielded a mixture of the diglucuronide and bilirubin, which suggested that pigment I was a 1/1 complex of the two. Attempts to reconstitute pigment I from the supposed constituents, pigment II and bilirubin, have usually failed but Nosslin[72] and Weber *et al.*[73] claimed success. The existence of a monoglucuronide. has now been proved by Ostrow and Murphy[74] who purified pigments I and II by solvent distribution. They mixed [^{14}C] bilirubin with the monoglucuronide and after separation found that there was no exchange of the labelled bilirubin with the conjugated pigment I. The monoglucuronide formed only biliverdin glucuronide on oxidation. Finally, when labelled monoglucuronide was administered to Gunn rats (which cannot excrete bilirubin) all the label was excreted.

There is little pigment I in human bile.[74] In the enriched bile of rats which had been given bilirubin intravenously, in order to increase the amount of excreted conjugated bilirubin, up to two-thirds of the conjugated pigment is monoglucuronide.[69,74] Monoglucuronide is also formed by *in vitro* systems (p. 132). Van Roy and Heirwegh[52] employed reverse-phase column chromatography to identify the product of a microsomal system as pigment I and confirmed this by determining the ratio of azobilirubin to conjugated azobilirubin. There are now two chromatographic systems which can be used for analysis of conjugated bilirubins (in addition to reverse phase column chromatography).[75,76] Chemical synthesis of mono- and di-glucuronides has already been reported by Thompson and Hofmann.[77] As the synthetic diglucuronide differed from Ostrow and Murphy's[74] pure concentrate it may be an α-glucuronide or possibly not a C_1 ester.[77]

C. OTHER CONJUGATES

The existence of alkali-stable, direct-reacting pigments in human bile, sometimes in large amounts,[65] led to a search[78,79] for sulphate conjugates. In more recent studies sulphation of bilirubin has yielded three products, one of which appears to be a direct-reacting, alkali-stable, pigment which is rapidly oxidized to biliverdin.[80] Small amounts were reported to occur in human bile. Extension of this work would be useful.

More recently Kuenzle[81] in attempting to define the alkali-stable pigments of bile opened a new chapter on bile pigment conjugates. His first step was to chromatograph human bile concentrates on a reverse-phase column system similar to that of Billing, Cole and Lathe.[65] The bands of bilirubin pigments I and II were eluted, coupled with diazotized aniline and the azopigments were purified through 2 successive columns. The result was a series of polar bands which were rigorously examined by means of glycosidases, gas chromatography and mass spectrometry. Kuenzle concluded the azopigment B from pigment II had 5 main constituents each of which was formed by condensation of a disaccharide (consisting of one or more hexuronic acids) in ester linkage with the carboxyethyl of bilirubin (Fig. 11). Surprisingly Kuenzle could find no evidence for azobilirubin glucuronide. Both the study of Kuenzle[81] and that of Billing, Cole and Lathe[65] are open to the criticism that the overall yield of pigment was low (5-10%) and thus the final product was not necessarily typical of the starting materials.

A second approach to the problem has involved a more direct, 2 stage, technique. Heirwegh and his colleagues[82] prepared the azopigments of bile by coupling with diazotized ethyl anthranilate and chromatographed the azopigments on thin layer silica (Fig. 12). The bands were eluted and examined by mass spectrometry and gas chromatography. The main polar band from human

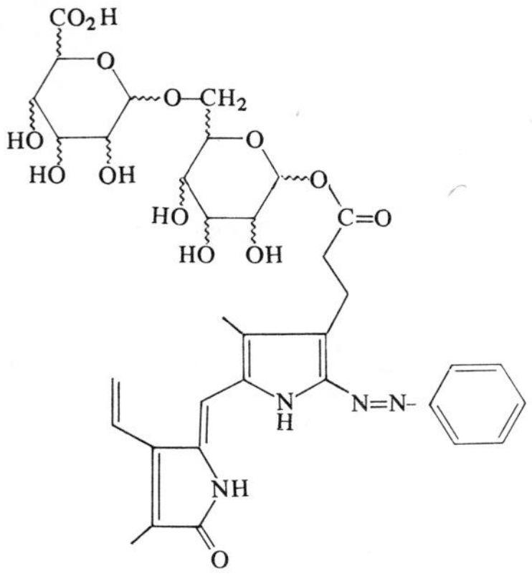

$B_{4\text{-}1}$

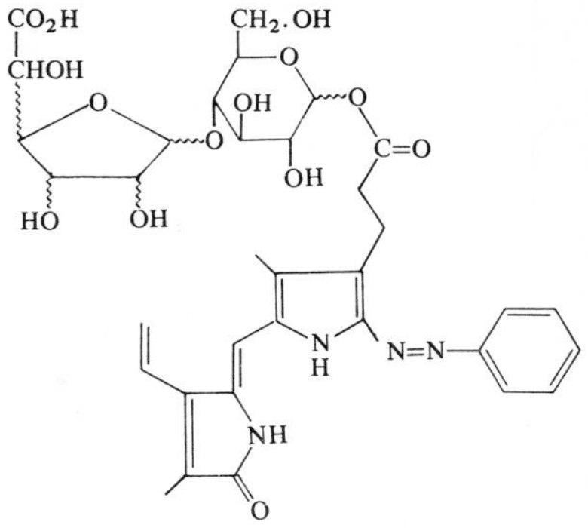

$B_{4\text{-}2}$

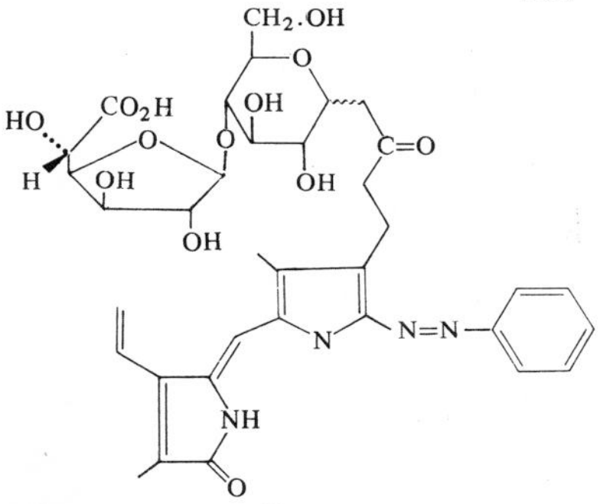

$B_{4\text{-}3}$

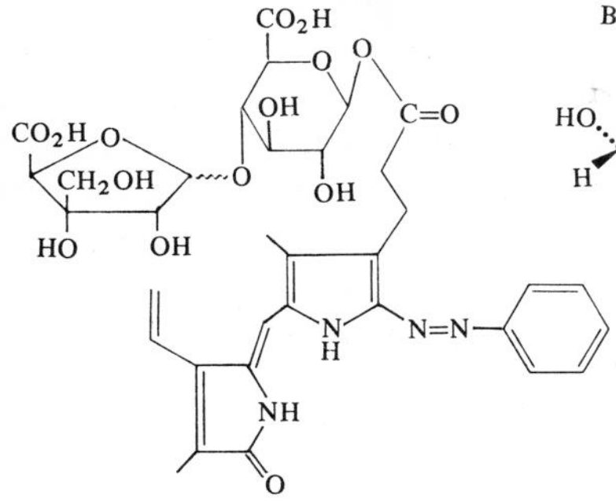

B_5

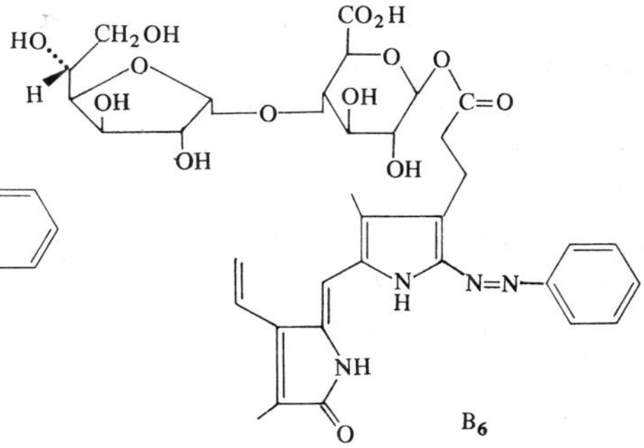

B_6

Fig. 11. Proposed structures of 5 of the conjugated azopigments isolated and partially characterized by Kuenzle.[81] They are: $B_{4\text{-}1}$ acyl 6-*O*-hexopyranosyluronic acid-hexopyranoside; $B_{4\text{-}2}$ acyl 4-*O*-hexofuranosyluronic acid-D-glucopyranoside; $B_{4\text{-}3}$ acyl 4-*O*-β-D-glucofuranosyluronic acid-D-glucopyranoside; B_5 acyl 4-*O*-(3-C-hydroxymethyl-D-ribofuranosyluronic acid)-β-D-glucopyranosiduronic acid; B_6 acyl 4-*O*-α-D-glucofuranosyl-β-D-glucopyranosiduronic acid.

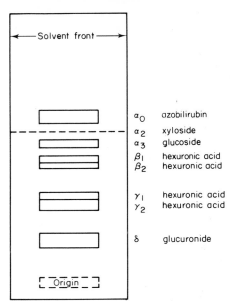

Fig. 12. Thin-layer chromatogram of a concentrate of rat bile which has been coupled with diazotized ethylanthranilate, according to Heirwegh *et al.*[82]

bile (δ, Fig. 12) was an alkali-labile, β-glucuronidase-labile, ester glucuronide. The most non-polar azopigment, α_0, was that of bilirubin. The intensity of the other bands varied with species; α_2 and α_3 occurred in trace amounts in human bile whereas in the dog they were major components.[83] They were the β-D-monoxyloside and β-D-monoglucoside respectively, in ester linkage.[49] The β_1 and β_2 pigments were not found in normal human bile but occurred after biliary obstruction. They and the γ_1 and γ_2 pigments were acid-labile, contained hexuronic acid, and were resistant to β-glucuronidase. The work of Kuenzle and of the Heirwegh group has opened up several new lines of promising investigation.

X. Formation of Conjugates

A. GLUCURONYL TRANSFERASES

Glucuronides of phenols, bilirubin, other carboxylic acid, some steroids[85] and thiols[84] are formed by microsomal, Mg^{2+} requiring, glucuronyl transferase (EC 2.4.1.17) as described by Dutton and Storey[85,86] in 1954-1955 (Fig. 13). Uridine diphosphate glucuronic acid (UDPGA), an important intermediate of carbohydrate metabolism, is the immediate source of the glucuronic acid

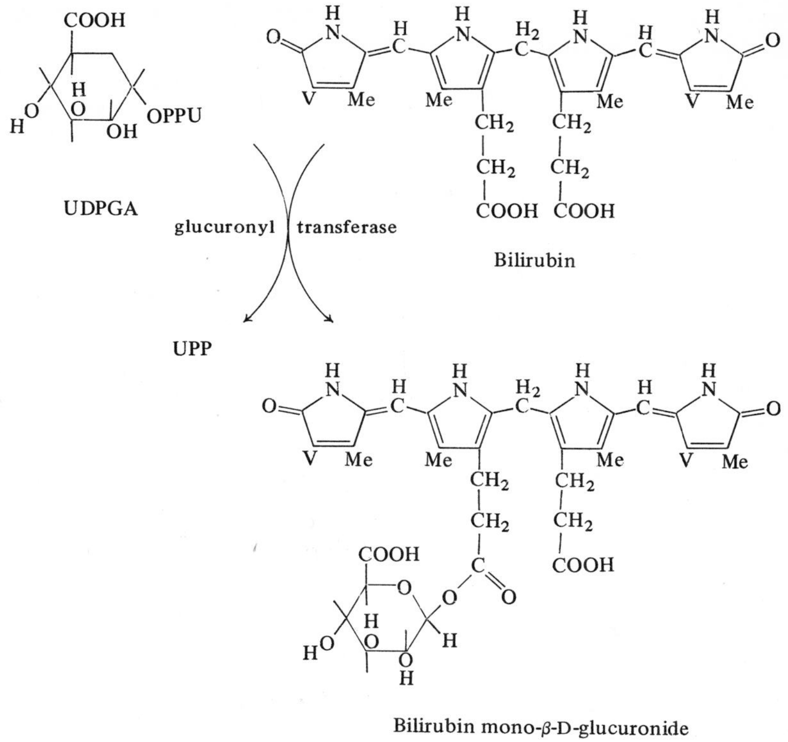

Bilirubin mono-β-D-glucuronide

Fig. 13. Reaction of bilirubin with UDPGA to form bilirubin mono-β-D-glucuronide. The diagram shows the transition from an α-substitution in UDPGA to the β-glucuronide.

residue. The enzyme is present in liver, kidney and to a lesser extent in gut and other tissues. The evidence for the existence of multiple enzymes, which has been reviewed by Dutton,[87] is based on differences in the relative activity with various substrates, the presence and absence of competitive inhibitions, the effect of non-competitive inhibitors, differences between species and between smooth and rough endoplasmic reticulum in the same species, variations in the loss of activities with heat, storage, salting out, or solvent treatment, differences in the rate of development of activity with different substrates, and differences in the effects of disease. The case for a separate enzyme for bilirubin is supported by the fact that man and the rat may have inherited defects affecting largely this substrate (p. 136). However, the bilirubin enzyme may also conjugate other aglycones.

White[88] found that the substrates bilirubin and *p*-nitrophenol were conjugated equally by rough and smooth endoplasmic reticulum unlike many other substrates for which smooth membrane is more active. The plasma membrane[88] and canalicular fractions of liver (author's unpublished studies) have no transferase activity. Halac et al.[89] reported that the smooth endoplasmic reticulum of rat liver synthesized only the monoglucuronide, while the rough reticulum synthesized both mono- and diglucuronide, and suggested that there were two enzymes.

Considerable attention has been given to the predominance of mono-glucuronide as the product of in vitro systems (p. 128), but this is consistent with there being one enzyme. It is unlikely that an enzyme would conjugate both carboxyls in one enzyme-substrate complex, as this would require two conjugating sites on the same protein. Alternatively the monoglucuronide may dissociate and then recombine with the enzyme site. In such a system bilirubin and the monoglucuronide which had been formed would compete for the enzymic site, and when bilirubin concentrations were high the formation of diglucuronide would be proportionately reduced. This appears to happen under two conditions of heavy bilirubin load: rats injected with bilirubin to increase the biliary concentration of conjugated pigment,[69,74] and in vitro systems operating at V_{max}.[52]

B. SOLUBILIZATION OF GLUCURONYL TRANSFERASE

Halac and Reff[90] solubilized microsomes with EDTA and deoxycholic acid and obtained a several-fold increase in activity with various substrates, including bilirubin, but Adlard and Lathe[91] later found that the same activity was obtained using deoxycholate alone. Heirwegh and co-workers[92,93] obtained partial solubilization and greatly enhanced activity with digitonin. Mowat and Arias[94] used ultrasonic oscillation, dialysis and ultracentrifugation ($4 \cdot 5 \times 10^6$ g-min). The apparently soluble material was fractionated on Sephadex G200 and the active fractions were examined by electronmicroscopy. Membranes were always found in the active fractions. These authors' method of solubilization and purification produced a greatly increased specific activity, but the total activity was not enhanced. Graham and Wood[95] degraded microsomes with phospholipases A and C. Although this left only 40% of the transferase activity (with p-nitrophenol as substrate) most of it was regenerated on addition of microsomal lipids.

The finding that some "solubilized" preparations show much higher activity may be taken as a priori evidence that this is the true estimate of activity in vivo but other explanations are possible, including conformational changes and the exposure of more sites. In untreated microsomal preparations the rate-limiting factor may be access of either substrate; the aglycone or UDPGA.

Pogell and Leloir[96] found that UDP-N-acetylglucosamine and ATP greatly increased the activity of microsomal preparations. Together the increase could be up to 25 fold. They could account for part of the increased activity by stabilization of UDPGA (through inhibition of UDPG pyrophosphatase). Adlard and Lathe[91] re-examined this with deoxycholate-treated enzyme. The pyrophosphatase had been inactivated and UDP-N-acetylglucosamine had no stimulating effect. ATP and a number of nucleotides activated the enzyme, probably by complexing divalent cations. Mg^{2+} activated the preparation but Ca^{2+} could substitute in its absence. At the optimum Mg^{2+} concentration Ca^{2+} inhibited, presumably by displacement.

Rat liver microsomes and solubilized preparations of bilirubin glucuronyl transferase have similar kinetic characteristics. K_m for bilirubin has been estimated to be 0·09 to 0·33 mmol/l and the K_m of UDPGA, 0·46 to 1·66 mmol/l.[97] Adlard and Lathe[91] found that the K_m for UDPGA was increased by some steroids.

Heirwegh and his colleagues[98] have prepared the glucoside and xyloside of bilirubin by incubating rat liver microsomes with UDP-glucose (UDPG) and UDP-xylose (UDPXyl). Digitonin increased the activity with UDPG and UDPXyl only slightly as compared with UDPGA. The transferase activity was greatest with UDPGA in the activated preparation and with UDPXyl using untreated microsomes. Wong[99] also found that the rate of formation of conjugated bilirubin (with EDTA-treated microsomes) was 4 times as great with UDPGA as with UDPG, even though the K_m of bilirubin with UDPG was half that with UDPGA. Microsomes do not form conjugated bilirubin with UDP-N-acetylglucosamine.[100] Heirwegh's group found that the Gunn rat was deficient in all three activities.[101] This would be consistent with the 3 activities being due to one enzyme having an affinity for the D-xylopyranose ring, which is the common feature of the three sugar nucleotides. Inhibitory studies with the three nucleotides should therefore prove interesting.

C. DEVELOPMENT OF GLUCURONYL TRANSFERASE

Glucuronyl transferase (acceptor : bilirubin) of the liver is very low at birth in the mouse,[102] rat,[103,100] guinea pig,[104] and rabbit[105] (Fig. 14). Activity with some phenolic substrates shows a different developmental pattern.[103] Examination of livers from human foetuses[106,107] or from infants dying immediately after delivery[108] reveals very low bilirubin conjugating activity. In the dog enzyme activity at birth is about half that of the adult.[109] Wong[99] found low glucosyl transferase activity in 6-day old rats.

The biological advantage of glucuronyl transferase being undeveloped *in utero* is not clear. Conjugation may not be necessary for disposing of bilirubin since it is a fat-soluble substance, probably easily cleared by the placenta. In fact, conjugated bilirubin, injected into the foetal circulation of guinea pigs and monkeys, is cleared more slowly than is bilirubin. There are marked species differences as shown by the accumulation of biliverdin in the dog placenta and of crystalline bilirubin in the seal placenta. There may, however, be a more general metabolic reason for the changes in glucuronyl transferase since it increases with hatching of chicks.[110]

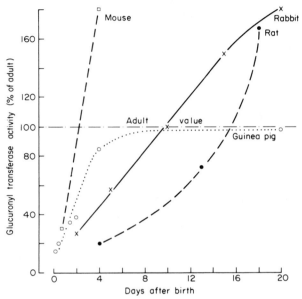

Fig. 14. Change in the activity of glucuronyl transferase (acceptor : bilirubin) of liver during the newborn period. Data have been plotted from determinations on homogenates of liver from the mouse,[102] guinea pig,[104] rabbit[105] and rat,[103] and are expressed in relation to the adult activity per gram liver.

Although the endoplasmic reticulum may be poorly developed at birth, this is not the reason for the low glucuronyl transferase activity as some microsomal enzymes are above adult values. Three possible explanations have been considered. Firstly, an enzyme inhibitor may be present. This is not supported by experiments in which homogenates of adult and newborn liver are mixed. Secondly, the initiation of enzyme protein synthesis may require a specific stimulus (acting presumably at the DNA, or possibly ribosome, stage). Thirdly, in the foetal stage there may be a specific inhibitor of transferase development which must be removed before enzyme is synthesized.

Greengard[111] has reviewed the evidence that many enzymes which develop during foetal, neonatal and weaning periods in the rat require some "initiator" or "trigger". Thus, glucocorticoids stimulate the increase in enzymes serving glycogenesis in the latter part of gestation and glucagon initiates the development of enzymes of gluconeogenesis. Few attempts have been made to initiate glucuronyl transferase during gestation. Flint et al.[105] gave growth hormone, insulin, secretin, hydrocortisone, tri-iodothyronine, bile salts and other substances to new born rabbits but did not succeed in increasing the rate of transferase development. Bakken and Fog[112,113] suggested that the substrate, bilirubin, was the initiating factor. Phenobarbitone, a potent inducer of endoplasmic reticulum in the adult, can enhance glucuronyl transferase slightly in the rat liver in the last few days of pregnancy.[103] This also occurs in the human (p. 138).

A more hopeful suggestion is that the structural gene for glucuronyl transferase is repressed *in utero* by a specific substance. Examples of specific substances which postpone metamorphosis are the juvenile hormone in the silk worm, and prolactin in the tadpole. Dutton and Ko[110] studied chick liver as a model of the development of glucuronyl transferase (substrate: *o*-aminophenol). It increased several-fold to adult levels in 2-4 h after hatching. Skea and Nemeth[114] maintained slices of liver on a grid in a humid atmosphere (organ culture) and found that the potential for development was expressed after day 5 and that enzyme synthesis appears to be repressed by some factor in the embryo environment. These studies suggest that an inhibitor should be looked for in mammals. Maternal administration of oestrogens can increase newborn hyperbilirubinaemia.[115]

D. PRODUCTION OF URIDINE DIPHOSPHATE GLUCURONIC ACID

The rate of glucuronide formation depends as much on the availability of the second substrate, UDPGA, as on bilirubin. UDPGA is produced by UDPG dehydrogenase (EC 1.1.1.22), a NAD-dependent, cytoplasmic enzyme, which transfers 4 reducing equivalents to 2 moles of NAD^+. UDPG dehydrogenase activity is low in the guinea-pig foetus and increases during the newborn period.[116] Flodgaard[117] found that there was a 30-fold increase in the concentration of UDPGA and a 100-fold increase in its turnover, between the foetal and adult state.

UDPG is an important branchpoint in carbohydrate metabolism. Uridine sugar nucleotides[99,118,119] are reported to be present in liver at concentrations of 0·1-1·4 m mol/l (except UDPXyl). UDPXyl, UDPG and UDPGA may be competing substrates for a transferase specific for xylopyranose, and UDPXyl[120] and UDPGal[121] may competitively inhibit UDPGA production. As UDPGA concentration is much less than the K_m, its concentration will affect the rate of conjugation *in vivo*. Starvation has little effect on the concentrations of UDPG and UDPGal,[122] but its effect on UDPGA has not been examined.

Dietary restrictions have been reported to increase plasma bilirubin in normal subjects, much more so in patients with Gilbert's disease (associated with a partial defect in glucuronyl transferase) and in some of their relatives,[123] in the Gunn rat[124] which lacks glucuronyl transferase (acceptor : bilirubin) and in the horse.[125] (A high fat diet lowers plasma bilirubin in the Gunn rat.[126])

Bakken et al.[34] reported that haem oxygenase was increased 2 to 3-fold by fasting in newborn and adult animals and suggested that a rise in plasma bilirubin may be due to increased bilirubin formation. This presupposes a pool of the precursor, haem, for which there is little evidence (p. 115). The effect of fasting on cytochrome P-450, which is a source of haem, is to depress the total amount in the liver but a possible change in the rate of turnover has not been examined.

XI. Inherited Defects of Bilirubin Conjugation

Three conditions have been described. The first, an apparently complete absence of the capacity to conjugate bilirubin, is represented by the Gunn strain[127,128,129] of Wistar rats and the Crigler-Najjar syndrome in man.[130,131] Both are inherited as autosomal recessive characters and in the homozygous an elevated plasma (unconjugated) bilirubin is present (in man 17-40 mg/100 ml) and the bile is free of pigment. The heterozygous rat has 55% of the enzyme activity of the normal.[100] The homozygous rat and human can form other glucuronides though some at a reduced rate. Fevery et al.[101] showed that the Gunn rat could not form the glucoside or xyloside.

The second condition, idiopathic constitutional hyperbilirubinaemia or Gilbert's disease, is associated with a plasma bilirubin of 1-3 mg/100 ml. It is probably inherited as an autosomal dominant.[132] Biopsy specimens conjugate 270 ± 140 (S.D.) μg bilirubin/g liver/h compared with the normal of 1100 ± 280.[132a] In many, if not all, of these patients the plasma bilirubin can be reduced by giving phenobarbitone.

In the third condition Arias's type II hyperbilirubinaemia[133] patients have a plasma bilirubin intermediate between that of patients with Gilbert's disease and the Crigler-Najjar syndrome. These patients have some conjugated pigment in their bile and respond to phenobarbitone, unlike Crigler-Najjar patients (type I of Arias). Inheritance is probably as an autosomal dominant.[133]

XII. Jaundice of the Newborn Infant

In newborn infants the plasma bilirubin concentration rises for a few hours or days (control group Fig. 15) and about 20% of infants become jaundiced with a plasma concentration of about 10 mg/100 ml or higher.[134] Small infants or those born prematurely require much longer for the peak to be reached and the mean peak value is much higher. As bilirubin is toxic the condition has been intensively studied.

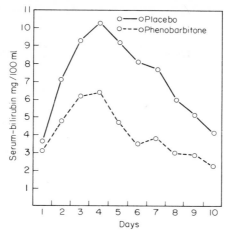

Fig. 15. The mean values of daily serum bilirubin determinations in a double blind study[141] of two groups of infants of normal weight and free of haemolytic disease. One group received intra-muscular phenobarbitone (10 injections of 5 mg at 8 h intervals) from birth. As the infants had been selected for hospital admission the values in the control (placebo) group are probably slightly higher than entirely unselected, normal weight, infants would have had.

A. GLUCURONYL TRANSFERASE DEFICIENCY

Accumulation of bilirubin is probably due primarily to a low glucuronyl transferase activity (p. 134).[134] Further, UDPGA may also be deficient, since UDPG dehydrogenase is also greatly lowered in some animals, e.g. guinea pig.[116] Other factors may also contribute: inhibition of glucuronyl transferase, reduced uptake of bilirubin by the liver due to undeveloped ligandin[64] (p. 125) and recirculation of bilirubin from the gut.

Lathe and Walker[135] noted that sera from pregnant women inhibited the *in vitro* conjugation of bilirubin by rat liver slices. A number of steroids, for example $3\alpha,20\alpha$-pregnanediol glucuronide, inhibited appreciably at concentrations as low as 1 μmol/1. However, steroids are probably not the primary cause of newborn jaundice because they had much less effect in broken cell preparations in which the transferase defect of the newborn infant is clearly evident.

Inhibition by steroids has been suggested as the cause of prolonged hyperbilirubinaemia, occurring in breast-fed infants. Arias[136] noted that the offending breast milk inhibited a number of glucuronide-forming systems, and isolated an unusual steroid, the 20β-isomer of "normal" $3\alpha,20\alpha$-pregnanediol, from the milk. Pregnanediol-20β inhibited bilirubin glucuronide formation by liver from some laboratory animals but not by liver of *Macaca mulatta* or of man.[137] Hargreaves and Piper[138] pointed to differences between the behaviour of the inhibitor present in human milk and the steroid and have suggested that the *in vitro* inhibition may be due to other substances, possibly fatty acids.

Adlard and Lathe[91] observed that bilirubin glucuronide formation by "solubilized" transferase from human liver was inhibited by oestriol and some other steroids at concentrations of 50 μmol/1. A Lineweaver-Burke analysis showed an "apparent" competitive inhibition with the second substrate, UDPGA. In view of the marked difference between the structures of oestrogens and UDPGA they suggested that this was an allosteric inhibition. The inhibition would therefore be increased in the presence of the reduced amounts of UDPGA which may occur in newborn infants (p. 135).

Intestinal reabsorption of the pigment would further exaggerate the problem of bilirubin excretion. The newborn infant has yellow stools containing much unconjugated bilirubin,[139] probably formed by intestinal β-glucuronidase. As bilirubin is fat-soluble it can be reabsorbed (p. 139). Poland and Odell[140] reduced hyperbilirubinaemia by feeding agarose, which binds bilirubin and may prevent its absorption.

Many attempts have been made to change the pattern of rise and fall of plasma bilirubin in newborn infants by giving phenobarbitone, or other agents inducing enzymes of endoplasmic reticulum[141] (cf. Fig. 15). Phenobarbitone will increase the glucuronyl transferase activity of adult human liver by about 40%.[142] It is uncertain however, that the depression of plasma bilirubin is due to increase in glucuronyl transferase. It might act by stimulating an alternative pathway of bilirubin disposal.

B. KERNICTERUS–TOXICITY OF BILIRUBIN

Jaundice of the brain nuclei, called kernicterus, occurs in some infants who have died with marked jaundice due to unconjugated bilirubin. It does not occur in adult jaundice due to conjugated bilirubin. It rarely appears at plasma bilirubin concentrations below 20 mg/100 ml and is very frequent above 30 mg/100 ml.

Kernicterus also occurs in Crigler-Najjar syndrome[143] where the hyperbilirubinaemia is the result of inherited absence of glucuronyl transferase (p. 136). Attempts to produce kernicterus in adult animals have usually been defeated by their extraordinary capacity to conjugate bilirubin. Lucey *et al.*[144] administered bilirubin repeatedly to newborn monkeys (*M. mulatta*) and were able to maintain bilirubin concentrations between 25 and 35 mg/100 ml for a day or more. These monkeys developed brain damage typical of kernicterus only when anoxia and hyperbilirubinaemia were combined. In view of the greater immaturity of the brain in the newborn infant, than in the monkey, it is possible that hyperbilirubinaemia may be the prime cause of kernicterus but asphyxia and acidosis may predispose to damage.

Zetterstrom and Ernster[145] showed that bilirubin was a potent uncoupler of oxidative phosphorylation. Mustafa *et al.*[146,147] found differences in its effect on mitochondria of liver and brain. Oxygen consumption of liver mitochondria was doubled at low concentration of bilirubin, whilst high concentrations inhibited respiration, reduced the P : O ratio, and lowered respiratory control. Brain mitochondria showed only the inhibitory effects. These authors concluded that an energy-dependent ingress of ions and water, typical of the effects of "transport inducing agents" resulted from an interaction of bilirubin with mitochondrial lipid. Under suitable conditions the toxic effect of bilirubin occurs with 1 μmol/l. Albumin, which binds bilirubin, protects as long as the ratio is less than 1 : 1.[146,148] In view of its toxicity bilirubin is removed from infants by exchange transfusion or irradiation,[149] the photo-decomposition products being non-toxic to mitochondria.[150]

XIII. Why is Conjugation Required for Bilirubin Secretion?

The Gunn rat cannot conjugate bilirubin, nor excrete it. It can, however, excrete intravenously administered conjugated bilirubin; its impaired ability to excrete must therefore be caused by its inability to conjugate. Some bile pigments are excreted unconjugated. The biochemical advantage in excreting bilirubin as a conjugate may be its greater polarity. Small lipid-soluble molecules readily pass cell membranes, and if unconjugated bilirubin were excreted it might be reabsorbed from the small bile channels. Bilirubin is more rapidly absorbed from the gut[151,152] and the gall bladder[153] than are the conjugates. In this view conjugation is a means of trapping bilirubin in the gut. A second suggestion is that conjugation may be necessary to adapt bilirubin to the secretory mechanism. Lester and Klein[154] have pointed out that mesobilirubinogen and *i*-urobilin are excreted unconjugated by normal rats and can be excreted by the Gunn rat. Molecular models of both pigments have a flexible structure unlike bilirubin. This may allow them to conform to the carrier protein of a secretory pump. The only difference between (meso) bilirubin and (meso) bilirubinogen lies in the unsaturation of the a and c bridges in the latter. If the conformation of bilirubin is dictated by internal hydrogen bonding (p. 122), conjugation may serve to break this and allow a less constrained conformation. However, the poor secretion of bilirubin dimethylester[155] does not support this theory. There is little information on other bile pigments. Jirsa and Billing (unpublished experiment) found that synthetic taurobilirubin, which is also polar, is rapidly excreted. Injected biliverdin glucuronide is excreted in the bile unchanged,[74] unlike biliverdin which is reduced and conjugated before excretion.[156]

XIV. Bilirubin Glucuronide Secretion

There is no aspect of bilirubin metabolism about which we are so ignorant as the secretory mechanism for conjugated bilirubin. The maximum capacity to secrete bilirubin (bilirubin T_m) can be determined in rats with canulated bile ducts by infusing bilirubin to saturate the secretory system.[157] Saturation occurs at a plasma bilirubin concentration of 10-15 mg/100 ml. For comparison, the rates at which rat liver conjugates and excretes bilirubin are listed:

	Rate (μg/g liver h)
Normal bilirubin secretion	6
Maximum bilirubin secretion (T_m)	1110
Maximum bilirubin conjugation by slices	59
Maximum bilirubin conjugation by untreated homogenate	439
Maximum bilirubin conjugation by solubilized liver	1500-2200

When T_m is being determined the main pigment is monoglucuronide; the rate of conjugation is therefore overestimated. Nevertheless the capacity to conjugate and to secrete is 100 or more times the normal secretion. It is not yet clear whether the rate-limiting factor is uptake, conjugation or secretion. Arias found comparable rates of bilirubin secretion following injections of conjugated bilirubin and unconjugated bilirubin in the rat[104] and the guinea pig.[104] The homozygous Gunn rat excretes conjugated bilirubin at the normal rate. The heterozygous Gunn rat with half the normal capacity to conjugate excretes bilirubin at half the normal rate, conjugation apparently being rate-limiting.

Elucidation of the mechanism transferring conjugated bilirubin from the endoplasmic reticulum, where it is formed, across the canalicular wall to the bile remains a real challenge to the biochemist.

REFERENCES

1. Hoard, J. L. (1971). Stereochemistry of hemes and other metalloporphyrins. *Science* **174**, 1295-1302.
2. *Enzyme Nomenclature Recomendations (1964) of the International Union of Biochemistry* (1965) p. 19, Elsevier, Amsterdam.
3. Tenhunen, R., Marver, H. S. & Schmid, R. (1969). Microsomal heme oxygenase. Characterization of the enzyme. *J. Biol. Chem.* **244**, 6388-6394.
4. Jackson, A. H., Kenner, G. W., Budzikiewiez, H., Djerassi, C. & Wilson, J. M. (1967) Pyrroles and related compounds. X Mass spectrometry in structural and stereochemical problems—XC Mass spectra of linear di-, tri- and tetrapyrrolic compounds, *Tetrahedron* **23**, 603-32.
5. Nichol, A. W. & Morell, D. B. (1969). Studies on the isomeric composition of biliverdin and bilirubin by mass spectrometry. *Biochim. Biophys. Acta* **184**, 173-183.

6. Bunn, H. F. & Jandl, J. H. (1968). Exchange of heme among hemoglobins and between hemoglobin and albumin. *J. Biol. Chem.* **243**, 465-475.
7. Gray, C. H., Neuberger, A. & Sneath, P. H. A. (1950). Studies in congenital porphyria. 2. Incorporation of ^{15}N in the stercobilin in the normal and in the porphyric. *Biochem. J.* **47**, 87-92.
7a. Israels, L. G., Levitt, M., Novak, W., Foerster, J. & Zipursky, A. (1967). "The Early Bilirubin" In ref. 58, pp. 3-14.
8. Murray, R. K., Connell, G. E. & Pert, J. H. (1961). The role of haptoglobin in the clearance and distribution of extracorpuscular hemoglobin. *Blood* **17**, 45-53.
9. Ostrow, J. D., Jandl, J. H. & Schmid, R. (1962). The formation of bilirubin from hemoglobin *in vivo*. *J. Clin. Invest.* **41**, 1628-1637.
10. Neale, F. C., Aber, G. M. & Northam, B. E. (1958). The demonstration of intravascular haemolysis by means of serum paper electrophoresis and a modification of Schumm's reaction. *J. Clin. Path.* **11**, 206-219.
11. Hrkal, Z. & Muller-Eberhard, U. (1971). Partial characterization of the heme binding serum glycoproteins rabbit and human hemopexin. *Biochemistry* **10**, 1746-1750.
12. Muller-Eberhard, U., Bosman, C. & Liem, H. H. (1970). Tissue localization of the heme-hemopexin complex in the rabbit and the rat as studied by light microscopy with the use of radioisotopes. *J. Lab. Clin. Med.* **76**, 426-431.
13. Drabkin, D. L. (1971). Heme binding and transport—a spectrophotometric study of plasma glycoglobulin hemochromogens. *Proc. Nat. Acad. Sci. U.S.* **68**, 609-613.
14. Goldfischer, G., Novikoff, A. B., Albala, A. & Biempica, L. (1970). Haemoglobin uptake by rat hepatocytes and its breakdown within lysosomes. *J. Cell Biol.* **44**, 513-519.
15. Kornfeld, S., Chipman, B. & Brown, E. B. (1969). Intracellular catabolism of hemoglobin and iron dextran by the rat liver. *J. Lab. Clin. Med.* **73**, 181-193.
16. Bissell, D. M., Hammaker, L. & Schmid, R. (1971). Erythrocyte catabolism in the liver. *Abstracts of the 22nd Annual Meeting of the American Society for the Study of Liver Diseases*. 7.
17. Tenhunen, R., Marver, H. S. & Schmid, R. (1970). The enzymatic catabolism of hemoglobin: stimulation of microsomal heme oxygenase by hemin. *J. Lab. Clin. Med.* **75**, 410-421.
18. Schmid, R., Marver, H. S. & Hammaker, L. (1966). Enhanced formation of rapidly labelled bilirubin by phenobarbital: hepatic microsomal cytochromes as a possible source. *Biochem. Biophys. Res. Commun.* **24**, 319-328.
19. Lemberg, R. (1956). The chemical mechanism of bile pigment formation. *Rev. Pure Appl. Chem.* **6**, 1-23.
20. Lemberg, R. & Legge, J. W. (1949). *Haematin Compounds and Bile Pigments*, Interscience Publishers Inc., New York.
21. Nakajima, H. (1963). Studies on heme α-methenyl oxygenase II The isolation and characterization of the final reaction product, a possible precursor of biliverdin. *J. Biol. Chem.* **238**, 3797-3801.
22. Levin, E. Y. (1967). The conversion of protohemochrome to verdohemochrome with liver homogenates. *Biochim. Biophys. Acta* **136**, 155-158.

23. Murphy, R. F., Ó Eocha, C. & Ó Carra, P. (1967). The formation of verdohaemochrome from pyridine protohaemochrome by extracts of red algae and of liver. *Biochem. J.* **104**, 6C.
24. Colleran, E. & Ó Carra, P. (1970). Non-enzymic nature of the pyridine haemochrome-cleaving activity of mammalian tissue extracts. (Haem α-methenyl oxygenase). *Biochem. J.* **119**, 905-911.
25. Nakajima, O. & Gray, C. H. (1967). Studies on haem α-methenyl oxygenase. Isomeric structure of formylbiliverdin, a possible precursor of biliverdin. *Biochem. J.* **104**, 20-22.
26. Ó Carra, P. & Colleran, E. (1969). Haem catabolism and coupled oxidation of haemproteins. *FEBS Lett.* **5**, 295-298.
27. Rüdiger, W. (1969). In *Porphyrins and Related Compounds* (Goodwin, T. W., ed.), pp. 121-130, Academic Press, London and New York.
28. Ó Carra, P. & Colleran, E. (1970). Methine-bridge specificity of the coupled oxidation of myoglobin and haemoglobin with ascorbate. *Biochem. J.* **119**, 42P-43P.
29. Colleran, E. & Ó Carra, P. (1970). Specificity of biliverdin reductase. *Biochem. J.* **119**, 16P-17P.
30. Ó Carra, P. & Colleran, E. (1971). Properties and kinetics of biliverdin reductase. *Biochem. J.* **125**, 110P.
31. Tenhunen, R., Marver, H., Pimstone, N. R., Trager, W. F., Cooper, D. Y. & Schmid, R. (1972). Enzymatic degradation of heme: oxygenative cleavage requiring cytochrome P-450. *Biochemistry* **11**, 1716-1720.
32. Pimstone, N. R., Engel, P., Tenhunen, R., Seitz, P. T., Marver, H. S. & Schmid, R. (1971). Inducible heme oxygenase in the kidney: A model for the homeostatic control of haemoglobin catabolism. *J. Clin. Invest.* **50**, 2042-2050.
33. Pimstone, N. R., Tenhunen, R., Seitz, P. T., Marver, H. S. & Schmid, R. (1971). The enzymatic degradation of hemoglobin to bile pigments by macrophages. *J. exp. Med.* **133**, 1264-1281.
34. Bakken, A. F., Thaler, M. M. & Schmid, R. (1972). Metabolic regulation of heme catabolism and bilirubin production I. *J. Clin. Invest.* **51**, 530-536.
35. Kondo, T., Nicholson, D. C., Jackson, A. H. & Kenner, G. W. (1971). Isotopic studies of the conversion of oxophlorins and their ferrihaems into bile pigments in the rat. *Biochem.* **121**, 601-607.
36. Jackson, A. H., Kenner, G. W. & Smith, K. M. (1968). Pyrroles and related compounds Part XIV The structure and transformation of oxophlorins (oxyporphyrins). *J. Chem. Soc.* (C), 301-310.
37. Singleton, J. W. & Laster, L. (1965). Biliverdin reductase of guinea pig liver. *J. Biol. Chem.* **240**, 4780-4789.
38. Tenhunen, R., Ross, M. E., Marver, H. S. & Schmid, R. (1970). Reduced nicotinamide-adenine dinucleotide phosphate dependent biliverdin reductase: partial purification and characterization. *Biochemistry* **9**, 298-303.
39. Overbeek, J. Th. G., Vink, C. L. J. & Deenstra, H. (1955). The solubility of bilirubin. *Recl. Trav. Chim Pays-Bas Belg.* **74**, 81-84.
40. Fog, J. & Jellum, E. (1963). Structure of bilirubin. *Nature (London)* **198**, 88-89.
41. Hutchinson, D. W., Johnson, B. & Knell, A. J. (1971). Tautomerism and hydrogen bonding in bilirubin. *Biochem. J.* **123**, 483-484.
42. Ó Carra, P. (1962). Zinc complex salt formation by bilirubin and mesobilirubin. *Nature (London)* **195**, 899-900.

43. Hutchinson, D. W., Johnson, B. & Knell, A. J. (1972). Metal complexes of bilirubin in dipolar aprotic solvents. *Tetrahedron Lett.* (in Press).
44. Nichol, A. W. & Morell, D. B. (1969). Tautomerism and hydrogen bonding in bilirubin and biliverdin. *Biochim. Biophys. Acta* 177, 599-609.
45. Kuenzle, C. C. (1970) Bilirubin conjugates of human bile. Nuclear-magnetic-resonance, infrared and optical spectra of model compounds. *Biochem. J.* 119, 395-409.
46. Overbeek, J. Th. G., Vink, C. L. J. & Deenstra, H. (1955). Kinetics of the formation of azobilirubin. *Recl. Trav. Chim. Pays-Bas. Blg.* 74, 85-97.
47. Hutchinson, D. W. Johnson, B. & Knell, A. J. (1972). The reaction between bilirubin and aromatic diazo compounds. *Biochem. J.* 127, 907-908.
48. Jansen, F. H. & Stoll, M. S. (1971). Separation and structural analysis of vinyl and isovinyl-azobilirubin derivatives. *Biochem. J.* 125, 585-597.
49. Compernolle, F., Van Hees, G. P., Fevery, J. & Heirwegh, K. P. M. (1971). Mass-spectrometic structure elucidation of dog bile azopigments of the acyl glycosides of glucopyranose and xylopyranose. *Biochem. J.* 125, 811-819.
50. Billing, B. H., Haslam, R. & Wald, N. (1971). Bilirubin standards and the determination of bilirubin by manual and Technicon autoanalyzer methods. *Ann. Clin. Biochem.* 8, 21-30.
51. Van Roy, F. P., Meuwissen, J. A. T. P., De Meuter, F. & Heirwegh, K. P. M. (1971). Determination of bilirubin in liver homogenates and serum with diazotised p-iodoaniline. *Clin. Chim. Acta* 31, 109-118.
52. Van Roy, F. P. & Heirwegh, K. P. M. (1968). Determination of bilirubin glucuronide and assay of glucuronyltransferase with bilirubin as acceptor. *Biochem. J.* 107, 507-518.
53. Brodersen, R. & Bartels, P. (1969). Enzymatic oxidation of bilirubin. *Eur. J. Biochem.* 10, 468-473.
54. Jacobsen, J. (1969). Binding of bilirubin to human serum albumin— determination of the dissociation constants. *FEBS Lett.* 5, 112-114.
55. Berk, P. D., Bloomer, J. R., Howe, R. B. & Berlin, N. I. (1970). Constitutional hepatic dysfunction (Gilbert's syndrome) A new definition based on kinetic studies with unconjugated radiobilirubin. *Amer. J. Med.* 49, 296-305.
56. Haymaker, L. & Schmid, R. (1967). Interference with bile pigment uptake in the liver by flavaspidic acid. *Gastroenterology* 53, 31-37.
57. Berthelot, P. & Billing, B. H. (1966). Effect of bunamiodyl on hepatic uptake of sulphobromophthalein in the rat. *Amer. J. Physiol.* 211, 395-399.
58. Bouchier, I. A. D. & Billing, B. H., eds. (1967). *Bilirubin metabolism*, Blackwell, Oxford and Edinburgh.
59. Hargreaves, T. (1966). The effect of male fern extract on biliary secretion. *Brit. J. Pharmacol. Chemother.* 26, 34-40.
60. Levi, A. J., Gatmaitan, Z. & Arias, I. M. (1969). Two hepatic cytoplasmic protein fractions Y and Z and their possible role in the hepatic uptake of bilirubin, sulfobromophthalein and other anions. *J. Clin. Invest.* 48, 2156-2167.
61. Litwak, G., Ketterer, B. & Arias, I. M. (1971). Ligandin: an abundant liver protein which binds steroids, bilirubin, carcinogens and a number of exogenous anions. *Nature (London)* 234, 466.
62. Ketterer, B. (1972). Proteins that bind carcinogen metabolites. *Biochem. J.* 126, 3P.

63. Levine, R. I., Reyes, H., Levi, A. J., Gatmaitan, Z. & Arias, I. M. (1971). Phylogenetic study of organic anion transfer from plasma into the liver. *Nature (New Biology)* **231**, 277-279.

64. Levi, A. J., Gatmaitan, Z. & Arias, I. M. (1970). Deficiency of hepatic organic anion-binding protein, impaired organic uptake by liver and "physiologic" jaundice in newborn monkeys. *New Engl. J. Med.* **283**, 1136-1139.

65. Billing, B. H., Cole. P. G. & Lathe, G. H. (1957). The excretion of bilirubin as a diglucuronide giving the direct van den Bergh reaction. *Biochem. J.* **65**, 774-784.

65a. Lathe, G. H. (1954). The chemical pathology of bile pigments, Part I. The plasma bile pigments. *Biochem. Soc. Symposia*, **12**, pp. 34-45.

66. Schmid, R. (1957). The identification of "direct reacting" bilirubin as bilirubin glucuronide. *J. Biol. Chem.* **229**, 881-888.

67. Schachter, D. (1957). Nature of the glucuronide in direct-reacting bilirubin. *Science* **126**, 507-508.

68. Compernolle, F., Jansen, F. H. & Heirwegh, K. P. M. (1970). Mass-spectrometric study of the azopigments obtained from bile pigments with diazotized ethyl anthranilate. *Biochem. J.* **120**, 891-894.

69. Jansen, F. H., & Billing, B. H. (1971). The identification of mono-glucuronides of bilirubin in bile as amide derivatives. *Biochem. J.* **125**, 917-919.

70. Talefant, E. (1956). Properties and composition of the bile pigments giving a direct diazo reaction. *Nature (London)* **178**, 312.

71. Gregory, C. H. (1963). Studies of conjugated bilirubin. III Pigment I, a complex of conjugated and free bilirubin. *J. Lab. Clin. Med.* **61**, 917-925.

72. Nosslin, B. (1960). The direct diazo reaction of bile pigments in serum: experimental and clinical studies. *Scand. J. Clin. Lab. Invest.* **12**, Suppl. No. 49, 1-76.

73. Weber, A. Ph., Schalm, L. & Witmans, J. (1963). Bilirubin mono-glucuronide (Pigment 1): A complex. *Acta Med. Scand.* **173**, 19-24.

74. Ostrow, J. D. & Murphy, N. H. (1970). Isolation and properties of conjugated bilirubin in bile. *Biochem. J.* **120**, 311-327.

75. Noir, B. A., Garay, E. R. & Royer, M. (1965). Separation and properties of conjugated biliverdin. *Biochim. Biophys Acta* **100**, 403-410.

76. Thompson, R. P. H. & Hofmann, A. F. (1971). Separation of bilirubin and its conjugates by thin layer chromatography. *Clinica Chim. Acta* **35**, 517-520.

77. Thompson, R. P. H. & Hofmann, A. F. (1970). Direct chemical synthesis of a bilirubin diglucosiduronic acid. *Gastroenterology* **60**, 202.

78. Isselbacher, K. J. & McCarthy, E. A. (1958). Identification of a sulfate conjugate of bilirubin in bile. *Biochim. Biophys. Acta* **29**, 658-659.

79. Schoenfield, L. J., Bollman, J. L. & Hoffmann, H. N. II (1962). Sulphate and glucuronide conjugates of bilirubin in experimental liver injury. *J. Clin. Invest.* **41**, 133-140.

80. Noir, B. A., Grozman, R. J. & De Walz, A. T. (1966). Studies on bilirubin sulphate. *Biochim. Biophys. Acta* **117**, 297-304.

81. Kuenzle, C. C. (1970). Bilirubin conjugates of human bile. The excretion of bilirubin as the acyl glycosides of aldobiuronic acid, pseudoaldobiuronic acid and hexuronosylhexuronic acid, with a branched-chain hexuronic acid as one of the components of the hexuronosylhexuronide. *Biochem. J.* **119**, 411-435.

82. Heirwegh, K. P. M., Van Hees, G. P., Leroy, P., Van Roy, F. P. & Jansen, F. H. (1970). Heterogeneity of bile pigment conjugates as revealed by chromatography of their ethyl anthranilate azopigments. *Biochem. J.* **120**, 877-890.

83. Fevery, J., Van Hees, G. P., Leroy, P., Compernolle, F. & Heirwegh, K. P. M. (1971). Excretion in dog bile of glucose and xylose conjugates of bilirubin. *Biochem. J.* **125**, 803-10.

84. Illing, H. P. A. & Dutton, G. J. (1970). Observations on the biosynthesis of thioglucuronides and thioglucosides in vertebrates and molluscs. *Biochem. J.* **120**, 16P-17P.

85. Dutton, G. J. & Storey, I. D. E. (1954). Uridine compounds in glucuronic acid metabolism. 1. The formation of glucuronides in liver suspensions. *Biochem. J.* **57**, 275-283.

86. Storey, I. D. E. & Dutton, G. J. (1955). Uridine compounds in glucuronic acid metabolism. 2. The isolation and structure of uridine diphosphate-glucuronic acid. *Biochem. J.* **59**, 279-288.

87. Dutton, G. J., ed. (1966). *The Biosynthesis of Glucuronides and Glucuronic Acid.* Academic Press, New York and London.

88. White, A. E. (1967). *The distribution of glucuronyl transferase in cell membranes*, in Bouchier I. A. D. & Billing B. H. (Ref. 58) p. 183-187.

89. Halac, E. Jr., Detwiler, P. & Dipiazza, M. (1970). The *in vitro* formation of bilirubin mono and diglucuronide. *Bull. N.Y. Acad. Med.* **46**, 460-461.

90. Halec, E. & Reff, A. (1967). Studies on bilirubin UDP-glycuronyl transferase. *Biochim. Biophys. Acta* **139**, 328-343.

91. Adlard, B. P. F. & Lathe, G. H. (1970). The effect of steroids and nucleotides on solubilized bilirubin uridine diphosphate-glucuronyl-transferase. *Biochem. J.* **119**, 437-445.

92. Heirwegh, K. P. M. & Meuwissen, J. A. T. P. (1968). Activation *in vitro* and solubilization of glucuronyltransferase (assayed with bilirubin as acceptor) with digitonin. *Biochem. J.* **110**, 31P-32P.

93. Black, M., Billing, B. H. & Heirwegh, K. P. M. (1970). Determination of bilirubin UDP-glucuronyltransferase activity in needle-biopsy specimens of human liver. *Clin. Chim. Acta* **29**, 27-35.

94. Mowat, A. P. & Arias, I. M. (1970). Partial purification of hepatic UDP-glucuronyltransferase. Studies of some of its properties. *Biochim. Biophys. Acta* **212**, 65-78.

95. Graham, A. B. & Wood, G. C. (1969). The phospholipid dependence of UDP-glucuronyltransferase. *Biochem. Biophys. Res. Commun.* **37**, 567-575.

96. Pogell, B. M. & Leloir, L. F. (1961). Nucleotide activation of liver microsomal glucuronidation. *J. biol. Chem.* **236**, 293-298.

97. Wong, K. P. (1971). Bilirubin glucuronyltransferase. Specific assay and kinetic studies. *Biochem. J.* **125**, 27-35.

98. Heirwegh, K. P. M., Meuwissen, J. A. T. P. & Fevery, J. (1971). Enzyme formation of β-D-monoglucuronide, β-D-monoglucoside and mixtures of β-D-monoxyloside, β-D-dixyloside of bilirubin by microsomal preparations from rat liver. *Biochem. J.* **125**, 28P-29P.

99. Wong, K. P. (1971). Formation of bilirubin glucoside. *Biochem. J.* **125**, 929-934.

100. Strebel, L. & Odell, G. B. (1971). Bilirubin uridine diphospho-glucuronyl-transferase in rat microsomes: genetic variation and maturation. *Pediat. Res.* **5**, 548-559.

101. Fevery, J., Heirwegh, K. P. M., Compernolle, F. & De Groote, J. (1971). Enzymic formation of bilirubin glucoside and bilirubinoxyloside, *6th meeting of the European Association for the Study of Liver*, Abstract 7, Eyre and Spottiswood.
102. Catz, C. & Yaffe, S. J. (1968). Barbiturate enhancement of bilirubin conjugation and excretion in young and adult animals. *Pediat. Res.* **2**, 361-370.
103. Halac, E. & Sicignano, C. (1969). Re-evaluation of the influence of sex, age, pregnancy, and phenobarbital on the activity of UDP-glucuronyl transferase in rat liver. *J. Lab. Clin. Med.* **73**, 677-685.
104. Gartner, L. M. & Arias, I. M. (1969). The transfer of bilirubin from blood to bile in the neonatal guinea pig. *Pediat. Res.* **3**, 171-180.
105. Flint, M., Lathe, G. H. & Ricketts, T. R. (1963). The effect of under nutrition and other factors on the development of glucuronyltransferase activity in the newborn rabbit. *Ann. N.Y. Acad. Sci.* **111**, 295-301.
106. Dutton, G. I. (1959). Glucuronide synthesis in foetal liver and other tissues. *Biochem. J.* **71**, 141-148.
107. Lucey, J. F. & Villee, C. A. (1962). Observations on human foetal hepatic UDPG dehydrogenase and glucuronyl transferase activity. *Abstracts of Papers, X International Congress of Pediatrics, Lisbon*; communication No 537.
108. Lathe, G. H. & Walker, M. J. (1958). The synthesis of bilirubin glucuronide in animal and human liver. *Biochem. J.* **70**, 705-712.
109. Lester, R. & Troxler, R. F. (1969). Progress in gastroenterology. Recent advances in bile pigment metabolism. *Gastroenterology* **56**, 143-169.
110. Dutton, G. J. & Ko, V. (1966). The synthesis of *o*-aminophenyl glucuronide in several tissues of the domestic fowl, *Gallus gallus*, during development. *Biochem. J.* **99**, 550-556.
111. Greengard, O. (1971). Enzymic differentiation in mammalian tissues. *Essays in Biochemistry* **7**, 157-205.
112. Bakken, A. F. & Fog. J. (1967). Bilirubin conjugation in newborn rats. *Lancet* **(2)** 309-310.
113. Bakken, A. F. (1969). Effect of unconjugated bilirubin on UDP-glucuronyl transferase activity in livers of newborn rats. *Pediat. Res.* **3**, 205-209.
114. Skea, B. R. & Nemeth, A. M. (1969). Factors influencing premature induction of UDP glucuronyl transferase activity in cultured chick embryo liver cells. *Proc. Nat. Acad. Sci. U.S.* **64**, 795-802.
115. Koivisto, K., Ojala, A. & Järvinen, P. A. (1970). The effect on neonatal bilirubin levels of oestrogen given to the mother before delivery. *Ann. Clin. Res.* **2**, 204-208.
116. Brown, A. K. & Zuelzer, W. W. (1958). Studies on neonatal development of glucuronide conjugating system. *J. Clin. Invest.* **37**, 332-340.
117. Flodgaard, H. (1968). UDPGA turnover in guinea-pig liver during perinatal development. *Abstracts of Fifth FEBS Meeting*, p. 104, Czechoslovak Biochemical Society, Praha.
118. Keppler, D. O. R., Rudigier, J. F. M., Bischoff, E. & Decker, K. F. A. (1970). The trapping of uridine phosphates by D-galactosamine, D-glucosamine and 2-deoxy-D-galactose A study on the mechanism of galactosamine hepatitis. *Eur. J. Biochem.* **17**, 246-253.
119. Keppler, D., Rudigier, J. & Decker, K. (1970). Enzymic determination of uracil nucleotides in tissues. *Analyt. Biochem.* **38**, 105-114.

120. Neufeld, E. F. & Hall, C. W. (1965). Inhibition of UDP-D-glucose dehydrogenase by UDP-D-xylose. A possible regulatory mechanism. *Biochem. Biophys. Res. Commun.* **19**, 456-461.
121. Hanninen, O. & Marniemi, J. (1970). Inhibition of glucuronide synthesis by physiological metabolites in liver slices. *FEBS Lett.* **6**, 177-181.
122. Keppler, D., Rudigier, J. & Decker, K. (1970). Trapping of uridine phosphates by D-galactose in ethanol-treated liver. *FEBS Lett.* **11**, 193-196.
123. Felsher, B. F., Pickard, D. & Redeker, A. G. (1970). The reciprocal relation between caloric intake and the degree of hyperbilirubinemia in Gilbert's syndrome. *New Engl. J. Med.* **283**, 170-172
124. Barrett, P. V. D. (1971). The effect of diet and fasting on the serum bilirubin concentration in the rat. *Gastroenterology* **60**, 572-576.
125. Gronwall, R. R. & Mia, A. S. (1969). Effect of starvation on bilirubin metabolism in the horse. *Physiologist* **12**, 241.
126. Housset, E. Etienne, J. P., Petite, J. P. & Manchon, P. (1966). Interet d'un regime alimentaire defini en experimentation animale. *Annals Biol. Clin.* **24**, 771-786.
127. Axelrod, J., Schmid, R. & Hammaker, L. (1957). A biochemical lesion in congenital non-obstructive non-haemolytic jaundice. *Nature (London)* **180**, 1426-1427.
128. Carbone, J. V. & Gradsky, G. M. (1957). Constitutional non-hemolytic hyperbilirubinemia in the rat: defect of bilirubin conjugation. *Proc. Soc. Expl. Biol. Med.* **94**, 461-463.
129. Lathe, G. H. & Walker, M. (1957). An enzyme defect in human neonatal jaundice and in Gunn's strain of jaundiced rats. *Biochem. J.* **67**, 9P.
130. Schmid, R. (1972). Hyperbilirubinemia, in *The Metabolic Basis of Inherited Disease* (Stanbury, J. B., Wyngaarden, J. B. & Fredrickson, D. S., eds.) 3rd. edn., p. 1141, McGraw Hill, New York.
131. Crigler, J. F. & Najjar, V. A. (1952). Congenital familial non-hemolytic jaundice with kernicterus. *Pediatrics* **10**, 169-180.
132. Powell, L. W., Hemingway, E., Billing, B. H. & Sherlock, S. (1967). Idiopathic unconjugated hyperbilirubinemia (Gilbert's syndrome) *New Engl. J. Med.* **277**, 1108-1112.
132a. Black, M. & Billing, B. H. (1969). Hepatic bilirubin UDP-glucuronyl transferase activity in liver disease and Gilbert's syndrome. *New Engl. J. Med.* **280**, 1266-1271.
133. Arias, I. M., Gartner, L. M., Cohen, M., Ezzer, J. B. & Levi, A. J. (1969). Chronic nonhemolytic unconjugated hyperbilirubinemia with glucuronyl transferase deficiency. *Am. J. Med.* **47**, 395-409.
134. Lathe, G. H. (1972). Liver function in the newborn. Symp. on "Liver Disease". Roy. Coll. Physicians Edin. (In press).
135. Lathe, G. H. & Walker, M. W. (1958). Inhibition of bilirubin conjugation in rat liver slices by human pregnancy and neonatal serum and steroids. *Quart. J. Exp. Physiol.* **43**, 257-265.
136. Arias, I. M., Gartner, L. M., Seifter, S. & Furman, M. (1964). Prolonged neonatal unconjugated hyperbilirubinemia associated with breast feeding and a steroid, pregnane-3 (alpha), 20 (beta)-diol, in maternal milk that inhibits glucuronide formation *in vitro*. *J. Clin. Invest.* **43**, 2037-2047.
137. Adlard, B. P. F. & Lathe, G. H. (1970). Breast milk jaundice: Effect of $3\alpha,20\beta$-pregnanediol on bilirubin conjugation by human liver. *Arch. Dis. Childh.* **45**, 186-189.

138. Hargreaves, T. & Piper, R. F. (1971). Breast milk jaundice. Effect of inhibitory breast milk and 3α,20β-pregnanediol on glucuronyl transferase. *Arch. Dis. Childh.* **46**, 195-198.
139. Broderson, R. & Hermann, L. S. (1963). Intestinal reabsorption of unconjugated bilirubin. A possible contributing factor in neonatal jaundice. *Lancet* (1) 1242.
140. Poland, R. L. & Odell, G. B. (1971). Physiologic jaundice: The enterohepatic circulation of bilirubin. *New Engl. J. Med.* **284**, 1-6.
141. Vest, M., Signer, E., Weisser, K. & Olafsson, A. (1970). A double blind study of the effect of phenobarbitone on neonatal hyperbilirubinaemia and frequency of exchange transfusion. *Acta Paediat. Stockh.* **59**, 681-684.
142. Billing, B. H. & Black, M. (1971). The action of drugs on bilirubin metabolism in man. *Ann. N.Y. Acad. Sci.* **179**, 403-410.
143. Lathe, G. H. (1964). Changing perspectives on bilirubin. *Proc. Ass. Clin. Biochem.* **3**, 103-109.
144. Lucey, J. F., Hibbard, E., Behrman, R. E., Esquivel de Gallardo, F. O. & Windle, W. F. (1964). Kernicterus in asphyxiated newborn rhesus monkeys. *Expl. Neurol.* **9**, 43-58.
145. Zetterstrom, R. & Ernster, L. (1956). Bilirubin, an uncoupler of oxidative phosphorylation in isolated mitochondria. *Nature (London)* **178**, 1335-1337.
146. Mustafa, M. G., Cowger, M. L. & King, T. E. (1969). Effects of bilirubin on mitochondrial reactions. *J. Biol. Chem.* **244**, 6403-6414.
147. Mustafa, M. G. & King, T. E. (1970). Binding of bilirubin with lipid A possible mechanism of its toxic reactions in mitochondria. *J. Biol. Chem.* **245**, 1084-1089.
148. Odell, G. B. (1970). The distribution and toxicity of bilirubin. *Pediatrics* **46**, 16-24.
149. Lucey, J. F. (1970). Phototherapy of Jaundice 1969 in *Bilirubin Metabolism in the Newborn* (D. Bergsma, ed.) Birth Defects 6, pp. 63-70. Williams and Wilkins Co., Baltimore.
150. Broughton, P. M. G., Rossiter, E. J. R., Warren, C. B. M., Goulis, G. & Lord, P. S. (1965). Effect of blue light on hyperbilirubinaemia. *Arch. Dis. Childh.* **40**, 666-671.
151. Lester, R. & Schmid, R. (1963). Intestinal absorption of bile pigments. I. The enterohepatic circulation of bilirubin in the rat. *J. Clin. Invest.* **42**, 736-746.
152. Lester, R. & Schmid, R. (1963). Intestinal absorption of bile pigments. II. Bilirubin absorption in man. *New Engl. J. Med.* **269**, 178-182.
153. Ostrow, J. D. (1967). Absorption of bile pigments by the gall bladder. *J. Clin. Invest.* **46**, 2035-2052.
154. Lester, R. & Klein, P. D. (1967). Why conjugation? Molecular structure and bilirubin glucuronide formation. In Bouchier & Billing (ref. 58), p. 89.
155. Jirsa, M., Dickinson, J. P. & Lathe, G. H. (1968). Effect on bilirubin excretion of blocking the carboxyl sites of glucuronide conjugation by methylation. *Nature (London)* **220**, 1322-1324.
156. Goldstein, G. W. & Lester, R. (1964). Reduction of biliverdin-C[14] to bilirubin-C[14] *in vivo. Proc. Soc. Expl. Biol. Med.* **117**, 681-683.
157. Weinbren, K. & Billing, B. H. (1956). Hepatic clearance of bilirubin as an index of cellular function in the regenerating rat liver. *Brit. J. Exp. Path.* **37**, 199-204.

Aldolase: A Model for Enzyme Structure-Function Relationships

C. Y. LAI* and B. L. HORECKER*

Department of Molecular Biology, Division of Biological Sciences, Albert Einstein College of Medicine, 1300 Morris Park Avenue, Bronx, N.Y. 10461, U.S.A.

I. Introduction

In 1907 Fletcher and Hopkins[1] made their classic discovery of the formation of lactic acid during the contraction of muscle and its disappearance during rest under aerobic, but not anaerobic, conditions. It was not until 1926, however, that Meyerhof[2] identified the source of this lactic acid as muscle glycogen; this marked the beginning of his classic experiments which led to the discovery[3] of aldolase, the enzyme which catalyses the cleavage of fructose 1,6-diphosphate to yield dihydroxyacetone phosphate and glyceraldehyde phosphate. This reaction had been postulated earlier by Embden and his coworkers[4] to account for the formation of 3-carbon compounds from hexose during glycolysis. The name aldolase was introduced when Meyerhof, Lohmann and Schuster[5,6] found that a variety of aldehydes, including both D- and L-glyceraldehyde 3-phosphate, and even acetaldehyde, would condense with dihydroxyacetone phosphate to form

*Present address: Roche Institute of Molecular Biology, Nutley, New Jersey 07110, U.S.A.

149

the corresponding aldol condensation products, according to the following scheme:

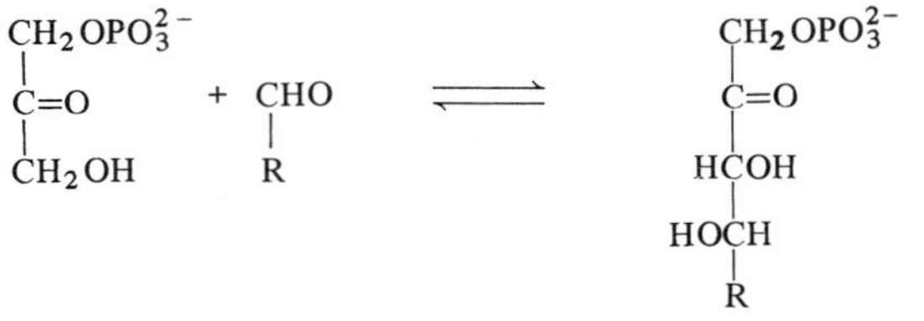

$$
\begin{array}{ccc}
\text{CH}_2\text{OPO}_3^{2-} & & \text{CH}_2\text{OPO}_3^{2-} \\
| & & | \\
\text{C=O} \quad + \quad \text{CHO} & \rightleftharpoons & \text{C=O} \\
| \qquad\qquad | & & | \\
\text{CH}_2\text{OH} \qquad \text{R} & & \text{HCOH} \\
& & | \\
& & \text{HOCH} \\
& & | \\
& & \text{R}
\end{array}
$$

Aldolase (EC 4.1.2.13) has been found to be widely distributed in nature, occurring in most microorganisms, and in nearly all plant and animal tissues (for a review see Horecker et al.[7]). In 1943 Warburg and Christian[8] first observed the presence of two distinct types of aldolases; the enzyme isolated from yeast was shown to require a metal ion for activity, unlike aldolase from muscle, which was not inhibited by chelating compounds such as cysteine, α,α'-dipyridyl or pyrophosphate. Later studies confirmed this fundamental difference between aldolases of higher plants and animals and those found in fungi and bacteria, and in 1964 these were formally classified by Rutter[9] as Class I and Class II aldolases, respectively.

Although aldolase was first crystallized from rat muscle by Warburg and Christian in 1943[8] it was not until 1948, when a simple procedure for crystallization of the enzyme from rabbit muscle was introduced by Taylor, Green and Cori,[10] that it became available on a large scale, making possible studies of its molecular properties and mechanism of action. However, even before this time a number of important clues as to the nature of the functional groups at the active site of the enzyme were available, largely as a result of the studies of Meyerhof and his collaborators on the specificity of the enzyme.[5,6] Thus they had established[5] that the condensation reaction was specific for dihydroxyacetone phosphate, while glyceraldehyde 3-phosphate could be replaced by a number of aldehydes, including simple aldehydes such as acetaldehyde or propionaldehyde. Furthermore, they showed that the enzyme-catalysed reaction was stereospecific,[6] yielding only products having the configuration of D-fructose, unlike the alkali-catalysed reaction studied earlier in the laboratories of Emil Fisher[11] and H. O. L. Fisher,[12] which produced D-sorbose in addition to D-fructose. This indication that the aldol condensation reaction involved a primary stereospecific activation of dihydroxyacetone phosphate was confirmed when it was established that the enzyme would catalyse the exchange of a proton of water with one of the 2 protons of the C-3 carbon atom of this substrate.[13] This led Bloom and Topper[14] to suggest the following mechanism for the enzyme-catalysed reaction:

$$
\begin{array}{c}
CH_2OPO_3H_2 \\
| \\
C=O \\
| \\
H-C-OH \\
| \\
H
\end{array}
\quad + En \quad \rightleftharpoons \quad H^+ \quad + \quad
\left[
\begin{array}{c}
CH_2OPO_3H_2 \\
| \\
C=O \\
| \\
H-C-OH \\
\ominus
\end{array}
\right] En \quad \longrightarrow
$$

$$
\left[
\begin{array}{c}
CH_2OPO_3H_2 \\
| \\
C-O^{\ominus} \\
\| \\
H-C-OH
\end{array}
\right] En
$$

and to propose that in the enzyme-substrate complex labilization of this proton was due to the stabilisation of an enolate ion. Indeed spectrophotometric evidence for the existence of an enzyme-dihydroxyacetone phosphate intermediate was obtained by Topper and his coworkers,[15] who observed absorption in the ultraviolet region of the spectrum which appeared when concentrated solutions of the enzyme were mixed with dihydroxyacetone phosphate. This complex was later characterized in our laboratory[16,17,18] as a Schiff base derivative involving dihydroxyacetone phosphate and a specific lysine residue at the active site of the enzyme. This Schiff base intermediate could be reduced with sodium borohydride, yielding a stable secondary amine which was isolated and characterized.[17,18] This discovery made it possible to label the active site of the enzyme and to carry out comparative studies on the structure of the active sites in aldolases derived from a variety of tissues and species. Additional methods, described below, have been employed to characterize other functional groups at the active site of rabbit muscle aldolase, and many of these functional groups, including the active site lysine residue, have been located in the primary amino acid sequence. This information has permitted some conclusions regarding the nature of the active site, as well as on the 3-dimensional folding of the molecule.

II. Rabbit Muscle Aldolase

A. MOLECULAR AND CATALYTIC PROPERTIES

1. Isolation and Characterization of the Enzyme

As was indicated earlier, crystalline rabbit muscle aldolase is readily obtained in large quantities by modification of the simple procedure described by Taylor et al.[10] in 1948. In this procedure the enzyme is precipitated from the crude muscle extract between 50% and 52% saturation with $(NH_4)_2SO_4$ at pH 7·5. The yield can be improved substantially by first removing the precipitate formed at 40% saturation with $(NH_4)_2SO_4$. After 2 recrystallizations the enzyme is essentially homogeneous by the criteria of electrophoretic mobility and ultracentrifugation. Contaminating traces of triosephosphate isomerase and glyceraldehyde 3-phosphate dehydrogenase can be removed by chromatography on DEAE-cellulose. The molecular weight of the enzyme has been estabilished as 158,000-160,000 by ultracentrifugation methods,[19,20] and this value agrees with that calculated from amino acid analysis.[21] At low pH, or in the presence of 8 M-urea or 6 M-guanidine HCl, the enzyme dissociates into subunits of approximately 40,000 molecular weight.[19,20,22,23] Activity lost during acid- or urea-induced dissociation can be recovered by dialysis or by dilution with buffer at neutral pH. Reassociation of subunits occurs, and virtually all of the molecular characteristics of the native enzyme are restored.[22,23]

Rabbit muscle aldolase catalyses the cleavage of fructose 1,6-diphosphate and sedoheptulose 1,7-diphosphate at nearly equal rates,[24] but the rate of cleavage of fructose 1-phosphate is nearly 50 times less than that of fructose diphosphate.[25,26] At 25°C 1 mg of pure aldolase catalyses the cleavage of 12-14 μmoles of fructose 1,6-diphosphate per minute. Fructose 6-phosphate does not serve as a substrate, confirming Meyerhof's early observation that the phosphate ester group at the C-1 position (corresponding to the phosphate group of dihydroxyacetone phosphate) is essential for catalytic activity, while the presence of the phosphate group at the C-6 position enhances the catalytic activity, but is not essential.[25-27]

2. Microheterogeneity and Subunit Structure

Investigators have noted for some time that the protein band formed in disc gel electrophoresis of crystalline rabbit muscle aldolase is rather broad. That this phenomenon is probably due to microheterogeneity of the enzyme was demonstrated by Susor et al.[28] when they succeeded in separating 5 active species by isoelectric focusing of the crystalline preparations. The source of this microheterogeneity appears to be the random association of 2 non-identical subunits to form the tetrameric structure.[29,30]

The presence of a multi-subunit structure and of non-identical subunits in rabbit muscle aldolase was first suggested by the experiments of Dreschler et al.[26] and Kowalsky and Boyer.[31] They found that treatment of rabbit muscle aldolase with carboxypeptidase A resulted in the release of 3 moles of tyrosine and 2 moles of alanine per mole of enzyme. This observation was confirmed by Winstead and Wold[32] with denatured aldolase, using a relatively large amount of carboxypeptidase A. It was suggested that the enzyme contained 2 different COOH-terminal sequences: -Ser-His-Ala-Tyr and -X-Ala-Tyr. Direct evidence for the presence of 2 non-identical subunits in rabbit muscle aldolase was obtained following their separation by Chan et al.[30] by chromatography on DEAE cellulose in 8 M urea. From the "α" chain carboxypeptidase A released Tyr,

Ala, His, and Ser, in that order, whereas similar treatment of the "β" chain yielded only Tyr and Ala.[33]

The carboxypeptidase digestion experiments mentioned in the preceding paragraph[26,27,31,32] which yielded 3 equivalents of tyrosine per mole of enzyme, as well as ultracentrifugation experiments,[22,23] indicated a 3-subunit structure for this enzyme. The 4-subunit structure, now widely accepted for rabbit muscle aldolase, was first suggested by the hybridization studies of Rutter and his coworkers,[34] and confirmed by the ultracentrifugation measurements of Kawahara and Tanford.[19] The enzyme was finally shown to yield 4 moles of tyrosine on digestion with carboxypeptidase after denaturing with urea[33] or maleic anhydride,[29] and 4 equivalents of an octadecapeptide were obtained following cleavage of the aldolase molecule with cyanogen bromide.[35] The separation of 5 bands in isoelectric focusing[28] suggested that the crystalline enzyme was composed of the 5 species: α_4, $\alpha_3\beta$, $\alpha_2\beta_2$, $\alpha\beta_3$, and β_4.

3. Origin of the α and β Subunits

Despite the evidence for 2 different polypeptide chains in rabbit muscle aldolase, recent results support the conclusion that there is only a single gene for this enzyme, and that the β subunit arises by modification of the α subunit. Koida et al.[29] found that only the α chain could be detected in aldolase isolated from rabbits younger than 3 months of age, and that the β chain appeared after this time, increasing to approximately the same amount as the α chain in mature animals (Fig. 1). This finding appeared to exclude a genetic basis for the microheterogeneity of rabbit muscle aldolase, and suggested a possible conversion of the α to the β subunit in vivo. The molecular basis for this conversion

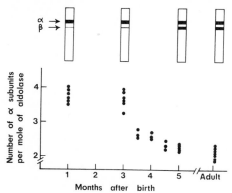

Fig. 1. Age-dependent appearance of β subunits in rabbit muscle aldolase. The upper portion of the figure shows the patterns in disc gel electrophoresis in 8 M-urea, and the lower portion the number of α subunits calculated from the ratio of histidine and tyrosine released by carboxypeptidase A.[29] Each point represents an aldolase preparation from a single animal.

appears to be the deamidation of a single asparagine residue in the α subunit. Thus the only difference thus far detected in the primary sequence of the α and β chains is in the fourth position from the COOH-terminus:[36]

Ile-Ser-*Asn*-His-Ala-Tyr (α chain)
Ile-Ser-*Asp*-His-Ala-Tyr (β chain)

This conversion of the α subunit to the β subunit through specific deamidation *in vivo* of an asparagine residue was first suggested by Lai and Horecker;[37] evidence to support this suggestion has recently been obtained by Midelfort and Mehler.[38] These workers injected rabbits with radioactive isoleucine and studied its incorporation into the α and β subunits. Label was shown to first enter the asparagine-containing peptide (α subunit) and only later was this chased into the aspartate-containing peptide (β subunit). They found a decreasing content of radioactivity in the 5 species of aldolase separated in isoelectric focusing, according to the following order: α_4, $\alpha_3\beta$, $\alpha_2\beta_2$, $\alpha\beta_3$, and β_4. The half-time for deamidation was found to be about 8 days, approximately equivalent to the half-life of the enzyme *in vivo*.[39] The mechanism of deamidation is as yet unknown; however, this *in vivo* modification of the enzyme does not appear to affect its catalytic activity.

4. *Primary and Tertiary Structure*

The NH_2-terminal amino acid of rabbit muscle aldolase was established by both chemical[40] and enzymic[41] methods to be proline. As already indicated, the COOH-terminal residue is tyrosine.[31] The protein contains 12 equivalents of methionine, or 3 per subunit, and the separation and ordering of the 4 cyanogen bromide peptides[21] has laid the groundwork for the recent studies on the primary amino acid sequence of the enzyme. The separation of the cyanogen bromide cleavage products on Sephadex G-75 is illustrated in Fig. 2. The largest peptide (peptide N), containing approximately 166 amino acid residues, is derived from the NH_2-terminal part of the molecule and contains the NH_2-terminal proline residue. The next largest peptide (peptide C), containing 122 amino acid residues, contains the tyrosine residue present at the COOH-terminus. This is followed by the peptide containing the active site lysine residue (peptide A), containing 66 amino acid residues. The final peptide to emerge from the column was peptide B, containing 18 amino acid residues. An overlapping tryptic peptide, containing the active site lysine residue and the adjacent methionine group, had already been isolated,[42] and this permitted the ordering of the A and B peptides, and therefore of all 4 of the cyanogen bromide peptides in the molecule.

The arrangement of the 4 cyanogen bromide segments and their approximate size is shown in Fig. 3. As indicated earlier, the presence of 12 methionine residues in the molecule and the isolation of 4 cyanogen bromide peptides

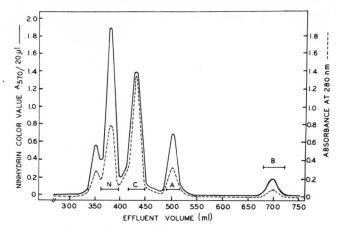

Fig. 2. Separation of the cyanogen bromide peptides of rabbit muscle aldolase. The chromatogram was obtained by gel filtration of 150 mg of aldolase, treated with BrCN, on a column of Sephadex G-75 (2.5 x 200 cm). The small peak eluted in front of peak N contained 10% of the total material and was found to be an aggregate of peptide fragments.[21]

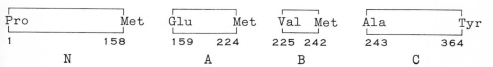

Fig. 3. Arrangement of the cyanogen bromide peptides in the rabbit muscle aldolase subunit.[7,21] The letters N, A, B, and C refer to the fractions isolated from the Sephadex G-75 columns (see Fig. 2).

confirmed the presence of 4 nearly identical subunits in the molecule,[21] and the recovery of 4 equivalents of the B peptide[35] provided additional evidence for this 4-subunit structure. Approximately two-thirds of the total amino acid sequence has now been determined (Fig. 4), and thus far no difference between the α and β subunits, other than the substitution of asparagine by aspartic acid discussed earlier, has been detected. It is of interest that the resistance of the β chain histidines to cleavage by carboxypeptidase persists even in the isolated COOH-terminal hexapeptides obtained by treatment of the molecule with chymotrypsin. Whereas the Asn-His linkage is susceptible to carboxypeptidase, the Asp-His sequence appears to be totally resistant.[36,46]

Of particular interest is the isolation and sequence analysis of the 8 cysteine-containing tryptic peptides in rabbit muscle aldolase. The primary sequence of these peptides and their appropriate positions are shown in Fig. 4. Although earlier studies had indicated that aldolase contained 28 cysteine residues per mole of enzyme, or 7 per subunit,[47] a total of 8 cysteine residues was later found by amino acid analyses of the BrCN peptides,[21] and the presence of 8 cysteine residues was also suggested by the later studies of Perham and Anderson.[45] This number for the cysteine peptides was confirmed by

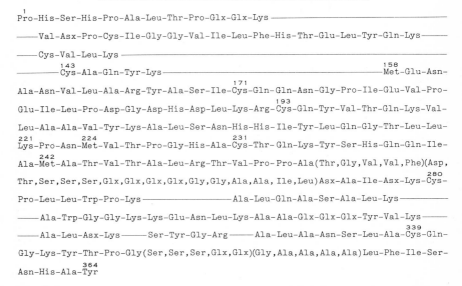

KNOWN AMINO ACID SEQUENCE IN RABBIT MUSCLE ALDOLASE

Fig. 4. The known sequence of amino acids in rabbit muscle aldolase. The data are from Lai (unpublished observations), Lai and Chen,[35] Lai et al.,[36,42] Lai and Oshima,[43] Sjago[44] and Perham and Anderson.[45]

Steinman and Richards[48] who used disulfide monosulfoxides, having the general structure $RCH_2CH_2S\text{—}S(=O)CH_2CH_2R'$. Experiments from a number of laboratories had suggested that there were 3 classes of thiol groups in rabbit muscle aldolase, based on the reactivity of these thiol groups toward a variety of reagents, including 5,5'-dithiobis-2-nitrobenzoate,[49] bromoacetate,[50] and iodo-acetamide,[51] as well as the disulfide monosulfoxides[48] already mentioned. These studies indicated the presence of 2-3 "exposed" thiol groups which reacted rapidly with thiol reagents either in the presence or absence of substrates, without loss of catalytic activity. Two additional thiol groups reacted less rapidly, and only in the absence of substrate. These were designated as "protected". Modification of these residues was associated with loss of catalytic activity. The remaining SH groups did not react with the reagents in the native protein, but only when it was denatured; these were considered to be "buried". The number and distribution of these sulfhydryl groups in the cyanogen bromide segments is illustrated in Fig. 5.

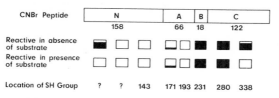

Fig. 5. Distribution and reactivities of sulfhydryl groups in rabbit muscle aldolase. The data are represented according to Steinman and Richards.[48] For the assignment of positions of cysteine residues see Fig. 4. The darkened areas represent reactivity of the cysteine residues.

Important information on the tertiary structure of the enzyme has been provided by the finding of a disulfide bridge which is formed under mild conditions of oxidation. It was first observed by Kobashi and Horecker[52] that the catalytic activity was lost when the enzyme was incubated at neutral pH in the presence of low concentrations of o-phenanthroline. The catalytic activity was quantitatively restored by the addition of reducing agents such as dithiothreitol or mercaptoethanol, and the loss of activity was shown to be due to the oxidation of 8 thiol groups, with the formation of one disulfide bridge per subunit. On the basis of the fact that only half of the activity was restored when the disulfide form of the enzyme was treated with cyanide, Kobashi and Horecker[52] concluded that one SH group was essential for the activity, while the other was located sufficiently close in the 3-dimensional structure to permit the ready formation of the disulfide bridge. The 2 thiol groups involved in disulfide formation have since been shown to be identical with the "protected" thiol groups and to be located at opposite ends of the molecule—one in the COOH-terminal segment at position 340, and the other in the NH$_2$-terminal segment, at a position yet to be determined.[53]

A schematic representation of the 3-dimensional structure of the molecule, based on the location and properties of the sulfhydryl groups, and particularly of the disulfide bridge, is shown in Fig. 6. The lysine residue, which forms the Schiff base with dihydroxyacetone phosphate, lies at position 221, near the center of the molecule. Proceeding from this lysine residue, the thiol groups located toward the COOH-terminal end of the chain appear to be exposed, on the surface of the molecule, whereas the thiol groups toward the NH$_2$-terminus from the active site lysine residue behave as though they were buried. However, both ends of the chain containing the "protected" thiol groups appear to fold back into proximity with the active site lysine residue. The tentative assignment of other functional groups in the primary sequence will be indicated in a later section. Recent evidence, based on fluorescence studies with alkyl or aryl phosphates, indicates that the active center is hydrophobic in character.[54]

It should be pointed out that the "buried" thiol groups may be involved in the interaction between subunits, rather than buried in the tertiary structure of the molecule. In any event, they are not readily accessible to sulfhydryl reagents when the enzyme is in the native tetrameric state.

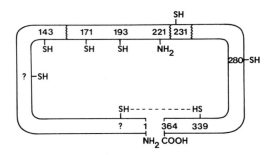

Fig. 6. Schematic diagram of the locations of the cysteine side chains and the active site lysine residue in rabbit muscle aldolase.

B. MECHANISM OF ACTION

1. *Formation of the Schiff Base Intermediate*

Although early workers had demonstrated the condensation of trioses in alkali, a specific catalytic role for primary and secondary amines was first indicated by studies of Westheimer and Cohen.[55] They showed that the dealdolization of dihydroxyacetone alcohol was specifically catalysed by primary and secondary amines, while tertiary amines were not effective. Since the rate of dealdolization increased with the amine concentration, they proposed that a ketimine intermediate was formed between the amine and the aldol substrate. Speck and Forist[56] first suggested that the catalysis of fructose 1,6-diphosphate cleavage by aldolase might be mediated by an amino group in the enzyme molecule which would combine with the carbonyl group of dihydroxyacetone phosphate.

Direct evidence for the formation of such a Schiff base intermediate was first reported by Horecker *et al.*[57] in studies with the related enzyme, transaldolase. A similar catalytic mechanism was soon discovered for rabbit muscle aldolase.[58] It was found that when aldolase was treated with $NaBH_4$ in the presence of ^{14}C or ^{32}P-labeled dihydroxyacetone phosphate, there was incorporation of radioactivity into the enzyme molecule, with concomitant loss of aldolase activity. After acid hydrolysis, the labeled amino acid derivative was isolated and identified as N^6-β-glycerolysine.[16,17] This was shown to be identical with the synthetic glycerolysine derivative,[18] and the same compound was isolated from the transaldolase-dihydroxyacetone complex after reduction with $NaBH_4$ and acid hydrolysis. It was suggested that the formation of the Schiff base intermediate and its reduction with $NaBH_4$ proceeded according to the following equations:

$$
\begin{array}{c}
H_2CO\textcircled{P}^{\ominus} \\
| \\
C=O \\
| \\
H_2COH
\end{array}
\ + \ H_2N{-}Lys \ \underset{-H_2O}{\rightleftharpoons} \
\begin{array}{c}
H_2CO\textcircled{P}^{\ominus}{}^{\oplus} \\
| \\
C=N{-}Lys \\
| \\
H_2COH
\end{array}
\ \rightleftharpoons \
\begin{array}{c}
H_2CO\textcircled{P}^{\ominus}{}^{\oplus} \\
| \\
C=N{-}Lys \\
| \quad H^{\oplus} \\
H_2COH
\end{array}
\ \xrightarrow{NaBH_4}
$$

$$
\begin{array}{c}
H_2CO\textcircled{P}^{\ominus} \\
| \quad H \\
HC{-}N{-}Lys \\
| \quad H^{\oplus} \\
H_2COH
\end{array}
$$

The obligatory formation of a Schiff base intermediate in the aldolase-catalysed reaction has been confirmed by the demonstration of oxygen exchange between the carbonyl group of fructose diphosphate and water, which is catalysed by the enzyme.[59] Additional evidence for the Schiff base intermediate has been reported, based on the formation of the cyanide addition product.[60]

The discovery of the Schiff base intermediate has not only provided evidence as to the mechanism of the enzyme-catalysed reaction, and facilitated studies on the structure of the active site of aldolase, but it has also provided an experimental basis for the classification of aldolases.[9] It has already been pointed out that Class I aldolases do not require metal ion cofactors and are not inhibited by chelating agents such as EDTA. These enzymes are inactivated by $NaBH_4$ treatment in the presence of the substrate dihydroxyacetone phosphate. Class II aldolases, on the other hand, are not inactivated by this treatment, but are inhibited by the addition of EDTA.[9]

2. Functional Groups at the Active Center

In addition to the active site lysine residue, a number of functional groups have been identified and shown to play a specific role in the reaction mechanism. In this section evidence for these groups is briefly documented, and where possible a role for these functional groups in the overall catalytic mechanism is suggested.

(a) *Lysine I.* We have already mentioned the lysine residue which forms a Schiff base intermediate with the carbonyl group of the substrate. Since only one such lysine residue is present in each enzyme subunit, its particular reactivity must be a result of the environment in which it is located. A tryptic peptide containing this lysine residue has been isolated and its sequence was reported;[42] similar sequences have since been found in a number of other Class I aldolases (see below).

In the case of the lysine residue in oxaloacetic acid decarboxylase, which forms a similar Schiff base with its substrate, Schmidt and Westheimer[61] have shown that the pK of this lysine residue is abnormally low, differing from the free ϵ-amino group of lysine by 4 pK units. Thus this lysine residue would not be protonated at physiological pH facilitating formation of the Schiff base intermediate. The abnormally low pK of this lysine residue may be attributed either to its presence in a hydrophobic environment, or, more likely, to the

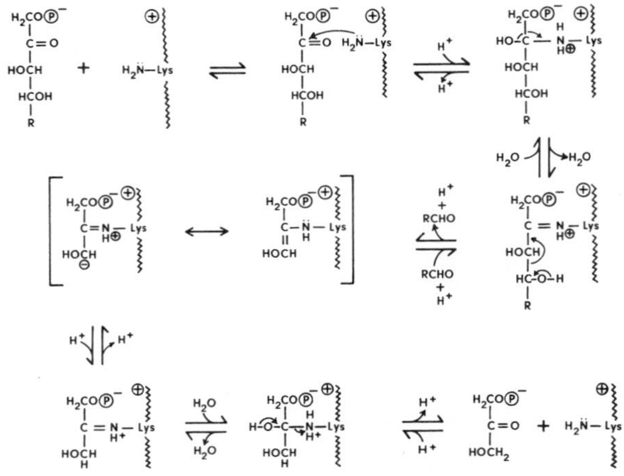

Fig. 7. Mechanism of dealdolization catalysed by rabbit muscle aldolase.

presence of an adjacent positively-charged group. The presence of such a positively-charged group in aldolase is indicated by the observations of Castellino and Barker[62] and Ginsburg and Mehler,[63] who identified a strong phosphate-binding site at the active center which is responsible for the interaction of the enzyme with the 1-phosphate group of the substrate. This high affinity phosphate-binding site is responsible for the primary interaction of the enzyme with the substrate, while at the same time it facilitates formation of the Schiff base intermediate. Ginsburg[64] was able to isolate this Schiff base intermediate without reduction with $NaBH_4$ and to show that both electrostatic interactions and Schiff base formation were involved in stabilization of the enzyme-substrate intermediate. Thus, the primary interaction of the substrate dihydroxyacetone phosphate with the enzyme may be represented as shown in Fig. 7. Schiff base formation would be favored by the electrostatic interaction between the negatively-charged phosphate group of the substrate and the as yet unidentified positively-charged group at the active center, which would also suppress the ionization of the neighboring ϵ-amino group. Following formation of the Schiff base intermediate, however, neutralization of this positive charge by the phosphate group would allow normal protonation of the Schiff base nitrogen, providing an electron sink which would facilitate the shift of electrons and labilization of the carbon-carbon bond. The first product of dealdolization would be expected to be the eneamine, a resonance form of the protonated carbanion (see Fig. 7).

(b) *Lysine II*. The presence of a second lysine residue at the active center, which may also play a role in the catalytic reaction, has been suggested by the work of Shapiro et al.,[65] using pyridoxal phosphate as a specific inhibitor of the enzyme. Inactivation by pyridoxal phosphate, which is reversible on dilution or on incubation with substrate, becomes irreversible when the inhibited enzyme is treated with $NaBH_4$, and the N^6-pyridoxyl derivative of lysine has been identified in the reduced product. Evidence that this lysine residue is distinct from that involved in the formation of the Schiff base intermediate with dihydroxyacetone phosphate was provided by the fact that the reduced N^6-pyridoxyl aldolase was still capable of forming the Schiff base derivative with the substrate.[65] The enzyme was also inactivated by free pyridoxal, but only at concentrations some 100-fold higher. It was suggested that the primary interaction of the inhibitor with the enzyme involved the high affinity phosphate binding site,[63] which positions the aldehyde group near the second lysine residue. According to this hypothesis the second lysine residue would function in the interaction of the enzyme with the C-6 phosphate group of fructose diphosphate. Examination of 3-dimensional models suggests that the distance between the 2 phosphate groups in the furanose form of fructose diphosphate is approximately equal to the distance from the aldehyde group to the phosphate group in pyridoxal phosphate.

From studies with model systems, Yasnikov et al.[66] have also proposed that a second basic group participates in the catalytic mechanism of aldolase, and specifically in the removal of the C-3 proton of dihydroxyacetone phosphate. In rabbit muscle aldolase this function appears to involve a histidine residue rather than lysine. The evidence for this is presented in the following section.

(c) *Histidines*. Aldolase becomes inactivated during irradiation in the presence of rose Bengal,[67] and this has been shown to be due to a specific destruction of histidine residues in the molecule. Inactivation was associated with loss of more than half of the 40 histidine residues in the molecule. However, loss of catalytic activity was incomplete, and the irradiated enzyme characteristically retained 10-20% of the original catalytic activity. Furthermore, it was still able to form the Schiff base intermediate with dihydroxyacetone phosphate. A most significant observation was that the catalytic activity in the presence of an added aldehyde acceptor was comparable to that of the native enzyme. Thus the ability of the enzyme to catalyse the transfer of dihydroxy-acetone phosphate from fructose diphosphate to an aldehyde acceptor was not impaired.[67] On the other hand, in the irradiated enzyme the ability to catalyse the labilization of a proton at the C-3 carbon atom of dihydroxyacetone phosphate and its exchange with tritium water was almost completely abolished. Based on these observations, it was proposed that the enzyme contained one or more essential histidine residues which catalysed the transport of protons to and from the Schiff base carbanion (Fig. 8). In the absence of this specific proton

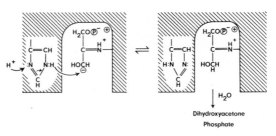

Fig. 8. Proposed role of histidine residues in the transport of protons to the Schiff base carbanion.

transport mechanism the release of the dihydroxyacetone phosphate moiety from the enzyme would be hampered, since protonation of the Schiff base intermediate is an essential requirement for this release (see Fig. 7). Thus, in the irradiated enzyme the rate-limiting step would be the release of the dihydroxy-acetone phosphate group from the Schiff base intermediate. The addition of a suitable aldehyde acceptor, such as acetaldehyde, would allow this group to be transferred, leaving the active site free to react with another molecule of fructose diphosphate. Thus, the photoinactivated enzyme behaves like transaldolase, which requires an aldehyde acceptor for release of the dihydroxyacetone residue[68] unlike aldolase, which is able to cleave the substrate to form 2 equivalents of triosephosphate.

It may be significant that the photoinactivation of aldolase in the presence of rose Bengal was found to proceed as rapidly at pH 5·5 as at pH 8·6.[67] Since protonated imidazole groups have been shown to be resistant to photooxidation in the presence of rose Bengal,[69] the abnormal behavior of the sensitive histidine residues in aldolase suggests that they are located in a hydrophobic environment. This observation is consistent with the suggestion (see above) that the active site forms a relatively hydrophobic pocket.

In the studies with rose Bengal as the photooxidant, where 5-6 histidine residues per subunit were oxidized, it could not be determined whether one or several histidine residues participated in the catalysis, nor was it possible to locate the essential histidine residues in the primary structure. A more specific photoinactivation of aldolase has been demonstrated using pyridoxal phosphate as the photosensitizer.[70,71] With this reagent, with either rabbit muscle aldolase or spinach aldolase, approximately one histidine per subunit is destroyed, yielding an enzyme with characteristics similar to those observed following photoinactivation with rose Bengal. Pyridoxal phosphate may indeed prove to be the reagent of choice for photoinactivation studies of aldolase and other enzymes.[72]

(d) *Tyrosines.* The identification of tyrosine as the COOH-terminal amino acid in the aldolase molecule has already been discussed. In 1959 Dreschler *et al.*[26] discovered that removal of these tyrosine residues by the action of carboxypeptidase A resulted in a dramatic decrease in the rate of cleavage of fructose 1,6-diphosphate, with little or no change in the rate of cleavage of fructose 1-phosphate. This observation was later confirmed by Rutter and his coworkers,[27] and Mehler and Cusic[73] showed that only V_{max} was affected by this treatment, while the K_m value for fructose diphosphate remained unchanged. These results suggested that the COOH-terminal residues might play a role in maintaining a conformation of the catalytic site which provides for an enhanced rate of cleavage of fructose diphosphate, as compared with fructose 1-phosphate. Mehler and Cusic also showed that the 5- and 8-carbon analogues of fructose diphosphate were cleaved at a rate comparable to that of fructose 1-phosphate, although the K_m values for these substrates were similar to that of fructose diphosphate. It has already been pointed out that sedoheptulose 1,7-diphosphate is cleaved at nearly the same rate as fructose diphosphate; thus either fructose 1,6-diphosphate or sedoheptulose 1,7-diphosphate provides the correct spacing between the 2 phosphate groups, permitting efficient interaction with the second phosphate-binding site (see below).

The enzyme which has lost the COOH-terminal tyrosine residues has other properties which resemble those of photooxidized aldolase. Release of the Schiff base carbanion becomes the rate-limiting step,[74] and the rate of formation of glyceraldehyde 3-phosphate is enhanced by the addition of an acceptor aldehyde.[75] A significant difference is that photoinactivation of the enzyme leads to a loss of catalytic activity toward both fructose diphosphate and fructose 1-phosphate, whereas removal of the COOH-tyrosine residues, as previously indicated, causes little change in the rate of fructose 1-phosphate cleavage.

Internal tyrosine residues in the aldolase molecule may play a role similar to that attributed to the COOH-terminal residues. Thus, similar changes in the catalytic properties of rabbit muscle aldolase are produced when the enzyme is treated with acetyl imidazole;[76,77] these changes include a loss in catalytic activity towards fructose diphosphate, with no change in the rate of cleavage of fructose 1-phosphate, and impairment of the ability to catalyse the proton exchange reaction. The acetylated enzyme also acts as a transaldolase and catalyses the transfer of the dihydroxyacetone phosphate group to aldehyde acceptors. The properties of the native enzyme are largely restored when the acetyl groups are removed by treatment with neutral hydroxylamine. During the treatment with acetyl imidazole more than half of the tyrosine residues in the enzyme are acetylated,[78] but the COOH-terminal tyrosine residues remain unmodified. The finding of similar properties when either internal or COOH-terminal tyrosine residues are modified suggests that these residues do not play a specific role in the catalytic mechanism, but rather that they are necessary to maintain the conformation of the active center.

(e) *Cysteines.* Early work on the modification of thiol groups in rabbit muscle adolase suggested that this enzyme was not a typical sulfhydryl enzyme. Studies by Swenson and Boyer[79] suggested that 8-10 thiol groups in the enzyme could be modified by *p*-mercuribenzoate without affecting the catalytic activity, but other workers[80] found later that 4-8 cysteine residues were essential for catalytic activity and were protected by the substrate (see above). We have already referred to the formation of the disulfide bridge by oxidation of the "protected" thiol groups. The suggestion that one of these thiol groups was essential for catalytic activity was supported by the observation that only 50% of the original activity was recovered when the disulfide bridge was reduced with cyanide, as compared with 100% reactivation by dithiothreitol, mercapto-ethanol, or cysteine.[52]

More direct evidence for the presence of a specific cysteine residue at the active center of the molecule was obtained by studies of the inactivation of the enzyme with its aldehyde substrates, glyceraldehyde 3-phosphate or erythrose 4-phosphate. Incubation of rabbit muscle aldolase with a small molar excess of erythrose 4-phosphate in the absence of the other substrate, dihydroxyacetone phosphate, caused a progressive loss of catalytic activity, associated with the incorporation of 4 equivalents of erythrose 4-phosphate per mole of enzyme and the loss of an equivalent number of titrable sulfhydryl groups.[81,82] These results suggested the involvement of a specific cysteine residue in the catalytic activity. A mechanism for this inactivation is proposed in the following section.

3. *A Model for the Catalytic Mechanism*

The Schiff base mechanism is consistent with the kinetic evidence obtained by Rose and his coworkers,[74] who have proposed an ordered sequence for the condensation reaction catalysed by rabbit muscle aldolases. The enzyme reacts first with dihydroxyacetone phosphate to form the primary enzyme-substrate complex. This complex then reacts with glyceraldehyde phosphate, followed by the release of fructose 1,6-diphosphate. The information relating to the functional groups and the properties of the modified protein which was reviewed in the previous section allows us to propose a tentative model for the structure of the active site and the role of these functional groups (Fig. 9). In the direction of dealdolization, the initial binding of the furanose form of fructose 1,6-diphosphate to the active site involves electrostatic forces between the C-1 phosphate residue of the substrate and the as yet undefined positively-charged group at the active center. We assume that this primary event involves this form of the substrate, since the free keto form accounts for less than 2% of the sugar in aqueous solution.[83] Formation of the enzyme-substrate complex is followed by addition of the ε-amino group of the enzyme to the carbonyl group of the substrate and the elimination of a molecule of water to form the Schiff base intermediate (see also Fig. 7). The elimination of water would be promoted by the negative charge of the adjacent phosphate group. Schiff base formation

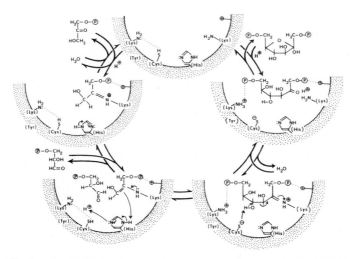

Fig. 9. A model for the catalytic mechanism of rabbit muscle aldolase.

would also be facilitated and the product stabilized by the hydrophobic environment around this lysine residue. This would be accompanied by protonation of the Schiff base nitrogen, which would draw electrons into the cationic electron sink, and labilize the bond between carbon atoms 3 and 4.

With fructose 1,6-diphosphate as the substrate, but not with fructose 1-phosphate, the rate of dealdolization would be further enhanced by withdrawal of a proton from the C-4 hydroxyl group. This is a result of the interaction of the C-6 phosphate group with the second lysine residue which enhances its basic properties and promotes the shift of a proton from a nearby thiol group to the lysine residue, with the formation of the thiolate ion, which would act as a conjugate base. This strongly basic group would promote the withdrawal of the proton from the C-4 hydroxyl group, shifting electrons into the C-4 carbon-oxygen bond and further labilizing the carbon-carbon bond.

Following dissociation of the glyceraldehyde moiety, protonation of the ketimine Schiff base, or more probably of the Schiff base carbanion, would be promoted by the histidine residue (or residues) in the active center (cf. Fig. 8). These histidine residues, which have been shown to have an abnormally low pK in the photoinactivation experiments with rose Bengal, may function as proton carriers and facilitate discharge of the carbanion and dissociation of the dihydroxyacetone phosphate moiety.

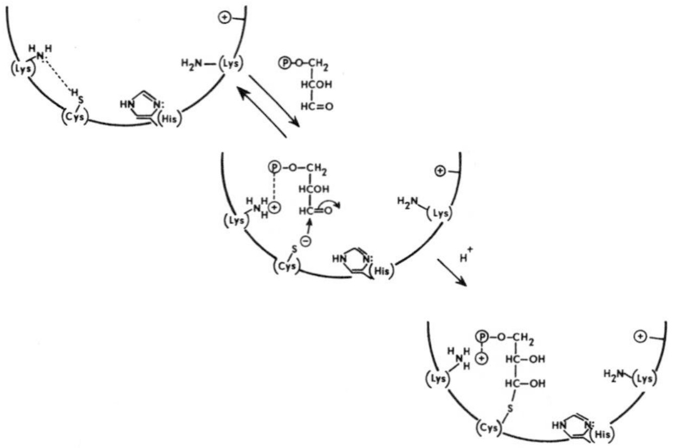

Fig. 10. The abortive reaction of aldehyde substrates with rabbit muscle aldolase.

Both the COOH-terminal tyrosine and internal tyrosine residues are seen as stabilizing the conformation which permits this specific interaction of the C-6 phosphate group with the second lysine residue.

The mechanism proposed also accounts for the inactivation of the enzyme when it is incubated with glyceraldehyde 3-phosphate or erythrose 4-phosphate in the absence of dihydroxyacetone phosphate. In the absence of the Schiff base carbanion, interaction of the phosphate group with the C-6 binding site would again generate the thiolate ion, which would now act as a strong nucleophile attacking the carbonyl group of the aldehyde substrate, and forming the hemithioacetal product (Fig. 10). This abortive reaction would impair or abolish the catalytic activity of the enzyme. It remains to be determined whether the thiol group in question is one of those involved in the formation of the disulfide bridge when the enzyme is inactivated with o-phenanthroline.

X-ray crystallographic measurements have confirmed the tetrameric structure of rabbit muscle aldolase,[49] but have thus far provided little information regarding the 3-dimensional structure of the subunit or the active site. However, a tentative model for the folding of the enzyme and the structure of the active site can be drawn from the information on the primary structure and the location of the functional groups. We have already discussed the evidence for folding of the molecule based on the reactivity of the sulfhydryl groups and the formation of the S-S bridge (Figs 5 and 6). The precise location of the protected sulfhydryl group in the N-peptide has not yet been determined, but we suggest that it may be rather remote from the active site lysine residue in the primary sequence.

Preliminary evidence suggests that the second lysine residue is also located in the N-peptide,[65,84] but information regarding the location of the essential histidine residue is not yet available.

III. Comparative Studies on Class I Aldolases

A. TISSUE-SPECIFIC ALDOLASES

1. *Occurrence of Aldolase Isoenzymes in Animal Tissues*

Three aldolase isoenzymes have now been identified in animal tissues. The presence of a distinct aldolase in mammalian liver was first reported by Leuthardt and his coworkers[85] and by Hers and Kusaka,[86] and the fact that this enzyme catalyzed the cleavage of fructose diphosphate and fructose 1-phosphate at nearly equal rates was established by Peanasky and Lardy.[87] Rutter and his coworkers[34,38] have identified 3 aldolase isoenzymes on the basis of their electrophoretic mobility, inhibition by specific antibodies, and the ratio of cleavage of fructose diphosphate and fructose 1-phosphate. These isoenzymes have been classified as aldolase A, the form found in muscle, aldolase B, the predominant form in liver and kidney, and aldolase C, which is found in brain together with aldolase A. Aldolase A characteristically shows lowest electrophoretic mobility and high specific activity towards fructose diphosphate. Aldolase B moves more rapidly toward the cathode at normal pH, and shows equal activity towards the 2 substrates. Aldolase C moves towards the anode in electrophoresis and shows a catalytic ratio of fructose diphosphate/fructose 1-phosphate which is intermediate between those of aldolases A and B. All 3 forms of aldolase form a Schiff base intermediate with dihydroxyacetone phosphate.

An extensive survey of aldolase isoenzymes, carried out by Lebherz and Rutter,[89] has established the presence of the 3 primary aldolase isoenzymes in a large number of animals, including fish, amphibia, birds and mammals.

2. *Molecular Properties and Subunit Structure*

Like aldolase A, aldolases B and C have tetrameric structures and are similar in molecular weight. The tetrameric structure was first demonstrated by Penhoet *et al.*[34] when they discovered that 5 forms of aldolase were present in brain tissue, and showed that the 3 species with intermediate electrophoretic mobilities were hybrids of aldolases A and C. Bands of similar mobilities were produced artifically on dissociation and reassociation of a mixture of these 2 aldolases *in vitro* (Fig. 11). Production of 5-membered sets, including 3 hybrid

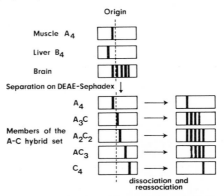

Fig. 11. Electrophoretic patterns of aldolase isoenzymes and the formation of the complete A-C hybrid sets on dissociation and reassociation of the separated hybrid isoenzymes. From Penhoet *et al.*[88]

forms, by reversible dissociation of mixtures of any of the 3 primary aldolases (A, B, or C) can only be explained if the parental aldolases were each composed of 4 identical subunits. Indeed this was the first evidence for the tetrameric structure of mammalian aldolases, later confirmed by ultracentrifugation and by the chemical methods previously cited.

Hybrid forms of aldolases have been found in certain tissues, suggesting the synthesis of subunits of more than one type of aldolase in the same cell,[34] since there is no evidence for the random exchange of subunits between aldolase tetramers under physiological conditions. Despite the tetrameric structure, the monomers in the hybrid forms appear to act independently without evidence of subunit interaction, and the catalytic properties of the heteropolymers have been shown to be identical to those of mixtures consisting of the corresponding proportions of homopolymers.[90,91] On the other hand, the presence of one subunit in a heteropolymer was sufficient to cause complete inactivation of aldolase on treatment with antibody against this subunit.[90] The active sites of the subunits in aldolase must therefore be physically close to each other in the tetrameric structure so that binding of specific antibody to one subunit will hinder the catalytic activity of the others.

3. *Comparison of the Active Site Structures*

Tryptic peptides containing the specific Schiff base-forming lysine residues have been isolated from aldolases A,[42] B,[92] and C,[93] and shown to have a striking degree of homology (Fig. 12), despite large differences in the amino acid composition of these enzymes.[93] All 3 enzymes show the same amino acid sequence in the region immediately surrounding the active site lysine residue, and only homologous substitutions can be detected in other parts of the peptide.

B. ACTIVE SITE PEPTIDES IN MUSCLE ALDOLASES FROM OTHER SPECIES

Aldolases have been isolated from the muscles of a variety of animal species, including mammals, birds, fish, amphibia, and invertebrates (for a review see Horecker *et al.*[7]). Active site peptides have been isolated from a number of these aldolases and their sequence determined (Fig. 12). The degree of homology among these muscle aldolases is remarkable. Except for isoleucine at position 7 in rabbit muscle aldolase, and aspartic acid and phenylalanine at positions 3 and 8 in lobster muscle aldolase, the sequences were found to be identical in the 20 amino acid residues forming this part of the active site peptide. Of particular interest is the uniform presence of the 2 histidine residues at positions 5 and 6 and the methionine residue at position 18. On the other hand, a histidine residue is found at position 23 only in the tryptic peptides from rabbit aldolases A and B. This suggests that this histidine residue is not essential for catalytic activity. Frog muscle contains only one methionine per subunit,[95] yet this methionine occupies the same position in the active site peptide. Also of interest is the exposed cysteine residue at position 25, which is present in all of the enzymes examined. This retention of structure is unexpected, particularly since the cysteine residue appears to be on the surface of the molecule and is not essential for catalytic activity. We have previously discussed the location of "buried" and "exposed" groups in the regions to the left and right of the active site lysine residue, respectively. It is noteworthy that amino acid replacements appear to be more frequent in the region containing the "exposed" cysteine residues, and less common where these are "buried". The left-hand part of the active site peptide is also rich in hydrophobic amino acid residues.

C. PLANT ALDOLASES

Plants have also been shown to contain Class I aldolases,[9] but to date only one active site peptide has been isolated and studied. A pentapeptide isolated from the active site of spinach leaf aldolase[98] has been shown to have the sequence Leu-Leu-Lys-Pro-(Ser). Except for the terminal serine residue which has not been identified with certainty, the sequence is identical to that found in

Source of aldolases	Residue Number 1 2 3 4 5 6 7 8 9 10 11 12 13 14 15 16 17 18 19 20 21 22 23 24 25 26 27 28
Rabbit A[42]	Ala-Leu-Ser-Asn-His-His-Ile-Tyr-Leu-Gln-Gly-Thr-Leu-Leu-Lys-Pro-Asn-Met-Val-Thr-Pro-Gly-His-Ala-Cys-Thr-Glu-Lys
B[92]	Ala-Leu-Asn-Asp-His-His-Val-Tyr-Leu-Glu-Gly-Thr-Leu-Leu-Lys-Pro-Asn-Met-Val-Thr-Ala-Gly-His-Ala-Cys-Thr-Lys-
C[93]	Ala-Leu-Ser(Asx,His,His,Ile,Tyr,Val,Glx,Ala,Thr,Leu,Leu,Lys,Pro,Glx)Met-Val-Thr-Pro-Gly-Asx-Ala-Cys-Thr-Gly-Lys
Codfish[94]	Ala-Leu-Ser-Asp-His-His-Val-Tyr-Leu-Gln-Gly-Thr-Leu-Leu-Lys-Pro-Asn-Met-Val-Thr-Ala-Gly-His-Ser-Cys-Ser-His-Lys
Frog[95]	Ala(Leu,Ser,Asx,His,His,Val,Tyr,Leu,Glx,Gly,Thr,Leu,Leu,Lys,Pro,Asx,Met)(Val,Thr,Ala,Gly,Asx,Ala,Cys,Thr,Glx)Lys
Sturgeon[96]	Ala-Leu-Ser-Asp-His-His-Val-Tyr-Leu-Gln-Gly-Thr-Leu-Leu-Lys-Pro-Asn-Met-Val-Thr-Ala-Gly-Gln-Ala-Cys-Thr-Lys-Lys
Lobster[97]	Ala-Leu-Asx-Asx-His-His-Val-Phe-Leu-Glx-Gly-Thr-Leu-Leu-Lys-Pro-Asn-Met-Val-Thr-Pro-Gly-Asx-Ala-Cys-Ser-Gly-Lys

Fig. 12. Comparison of the amino acid sequences of the active site peptides from Class I aldolases. The data are taken from the references indicated.

aldolases from animal tissues. It will be of particular interest to determine whether the next position is occupied by a methionine residue as in all other aldolases yet studied.

IV. Properties of Class II Aldolases

Although it is beyond the scope of this essay, it is of interest to compare some of the properties of yeast aldolase with those of rabbit muscle aldolase. Yeast aldolase is a dimer, having approximately half the molecular weight of the muscle enzyme and contains 2 equivalents of zinc per mole.[99,100]

A. MECHANISM OF ACTION

The metal ion at the active center of Class II aldolases is thought to catalyse the reaction by forming a partial bond with the carbonyl group of the substrate.[9] Thus the ternary complex formed between dihydroxyacetone phosphate, metal ion and the enzyme would be equivalent to the Schiff base intermediate formed in Class I aldolases. In the direction of condensation this would labilize the proton at the C-3 carbon atom and facilitate the nucleophilic attack on the aldehyde substrate as illustrated below.

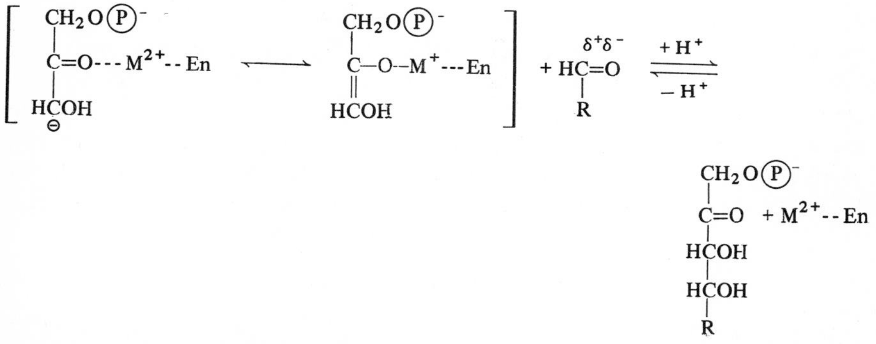

Direct interaction of the substrates with the two tightly bound manganese atoms in the Mn^{2+} form of yeast aldolase has recently been demonstrated by Mildvan et al.[101] using electron paramagnetic resonance methods. The metal ions appear not only to form a coordination bond with the carbonyl group of dihydroxyacetone phosphate, but also to serve as the C-1 phosphate binding site (Fig. 13).

Some studies have also been carried out on the role of other amino acid side chains in the catalytic action of metallo aldolases. The proton exchange at the C-3 carbon atom of dihydroxyacetone phosphate appears to be facilitated by the β-carboxyl group of an aspartic acid or γ-carboxyl group of a glutamic acid residue in the case of Class II aldolases from *Fusarium oxysporum.*[102] This is the role which has been proposed for histidine in the Class I aldolases.

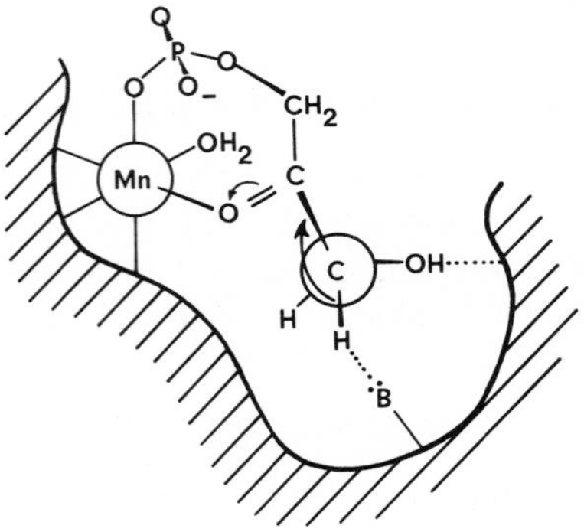

Fig. 13. A proposed mechanism of the action of yeast aldolase. From Mildvan *et al.*[101]

B. PRIMARY STRUCTURE

Although it was suggested,[100] on the basis of a computer analysis of the amino acid sequence of yeast and rabbit muscle aldolase, that there might be extensive homology in the primary structures of these two enzymes, such homology has not been found in any of the cysteine peptides. These have recently been isolated and sequenced by Harris *et al.*,[103] who found none of the cysteine-containing peptides of yeast aldolase to show any homology with the cysteine peptides of rabbit muscle aldolase. It would appear that the two classes of aldolases have arisen by convergent evolution, as suggested earlier by Rutter.[9]

V. Concluding Remarks

As is apparent from the above discussion, aldolase provides an excellent model for studies of structure-function relationships of an enzyme. Chemical modifications of amino acid side chains have been correlated with loss of activity, and in many cases it has been possible to relate these to changes in the rates of specific steps in the reaction sequence, thus providing evidence for the nature of the amino acids which are involved in the catalysis of these steps. The availablity of the pure protein in large quantities, and the existence of at least 3 isoenzymes in animal tissues, have provided additional parameters for the study of its catalytic activity. Similarly, because of its ubiquity throughout the plant and animal kingdom and the ease with which it can be prepared in pure form, aldolase has become a useful model for the analysis of phylogenetic relationships. The presence of tissue-specific aldolases also permits the use of this enzyme for studies of control of gene expression in higher organisms. Such studies are already underway in a number of laboratories.

Finally, aldolase serves as an excellent model for the analysis of subunit interactions, particularly since it is possible to assemble immunologically distinct subunits to form active hybrid enzymes.

REFERENCES

1. Fletcher, W. M. & Hopkins, F. Gowland. (1907). Lactic acid in amphibian muscle. *J. Physiol. (London)* **35**, 247-309.
2. Meyerhof, O. (1926). Über die Isolierung des glykolytischen Ferments aus dem Muskel und den Mechanismus der Milchsäurebildung in Lösung. *Naturwissenschaften* **14**, 1175-1180.
3. Meyerhof, O. & Lohmann, K. (1934). Über die enzymatische Gleichgewichtsreaktion zwischen Hexosediphosphorsäure und Dioxyacetonphosphorsäure. *Biochem. Z.* **271**, 89-110.
4. Embden, G., Deuticke, H. J. & Kraft, G. (1933). Über die Intermediaren Vorgänge bei der Glykolyse in der Muskulatur. *Klin. Wochenschr.* **12**, 213-215.
5. Meyerhof, O., Lohmann, K. & Schuster, Ph. (1936). Über die Aldolase, ein Kohlenstoff-verknüpfendes Ferment. II. Mitteilung: Aldolkondensation von Dioxyacetonphosphorsäure mit Acetaldehyd. *Biochem. Z.* **286**, 301-318.
6. Meyerhof, O., Lohmann, K. & Schuster, Ph. (1936). Über die Aldolase, ein Kohlenstoff-verknüpfendes Ferment. II. Mitteilung: Aldolkondensation von Dioxyacetonphosphorsäure mit Glycerinaldehyd. *Biochem. Z.* **286**, 219-335.
7. Horecker, B. L., Tsolas, O. & Lai, C. Y. (1972). Aldolases. In *The Enzymes*, (Boyer, P. D., ed.), 3rd ed. vol. 7, pp. 213-258, Academic Press, New York.
8. Warburg, O. & Christian, W. (1943). Isolierung und Kristallisation des Gärungsferments Zymohexase. *Biochem. Z.* **314**, 149-176.
9. Rutter, W. J. (1964). Evolution of aldolase. *Fed. Proc. Fed. Amer. Soc. Exp. Biol.* **23**, 1248-1257.
10. Taylor, J. F., Green, A. A. & Cori, G. T. (1948). Crystalline aldolase. *J. Biol. Chem.* **173**, 591-604.
11. Fischer, E. & Tafel, J. (1887). Synthetische Versuche in der Zuckergruppe. *Berlin Berichte* **20**, 2566-2575.
12. Fischer, H. O. L. & Baer, E. (1936). Synthese von *d*-Fructose und *d*-Sorbose aus *d*-Glycerinaldehyd, bzw. aus *d*-Glycerinaldehyd und Dioxyaceton; über Aceton-glycerinaldehyd III. *Helv. chim. Acta* **19**, 519-532.
13. Rose, I. A. & Rieder, S. V. (1955). The mechanism of action of muscle aldolase. *J. Amer. Chem. Soc.* **77**, 5764-5765.
14. Bloom, B. & Topper, Y. J. (1956). Mechanism of action of aldolase and phosphotriose isomerase. *Science* **124**, 982-983.
15. Topper, Y. J., Mehler, A. H. & Bloom, B. (1957). Spectrophotometric evidence for formation of a dihydroxyacetone phosphate-aldolase complex. *Science* **126**, 1287.

16. Grazi, E., Rowley, P. T., Cheng, T., Tchola, O. & Horecker, B. L. (1962). The mechanism of action of aldolases III. Schiff base formation with lysine. *Biochem. biophys. Res. Commun.* **9**, 38-43.

17. Horecker, B. L., Rowley, P. T., Grazi, E., Cheng, T. & Tchola, O. (1963). The mechanism of action of aldolases IV. Lysine as the substrate-binding site. *Biochem. Z.* **338**, 36-51.

18. Speck, Jr., J. C., Rowley, P. T. & Horecker, B. L. (1963). Identity of synthetic N^6-β-glyceryllysine and the C^{14}-labeled amino acid obtained on sodium borohydride reduction and hydrolysis of a complex from C^{14}-fructose 6-phosphate-transaldolase interaction. *J. Amer. Chem. Soc.* **85**, 1012-1013.

19. Kawahara, K. & Tanford, C. (1966). The number of polypeptide chains in rabbit muscle aldolase. *Biochemistry* **5**, 1578-1584.

20. Sia, C. L. & Horecker, B. L. (1968). The molecular weight of rabbit muscle aldolase and the properties of the subunits in acid solution. *Archs Biochem. Biophys.* **123**, 186-194.

21. Lai, C. Y. (1968). Studies on the structure of rabbit muscle aldolase I. Cleavage with cyanogen bromide: an approach to the determination of the total primary structure. *Archs Biochem. Biophys.* **128**, 202-211.

22. Deal, W. C., Rutter, W. J. & Van Holde, K. E. (1963). Reversible dissociation of aldolase into unfolded subunits. *Biochemistry* **2**, 246-251.

23. Stellwagen, E. & Schachman, H. K. (1962). The dissociation and reconstitution of aldolase. *Biochemistry* **1**, 1056-1068.

24. Horecker, B. L., Smyrniotis, P. S., Hiatt, H. H. & Marks, P. A. (1955). Tetrose phosphate and the formation of sedoheptulose diphosphate. *J. Biol. Chem.* **212**, 827-836.

25. Tung, T.-C., Ling, K.-H., Byrne, W. L. & Lardy, H. A. (1954). Substrate specificity of muscle aldolase. *Biochim. biophys. Acta* **14**, 488-494.

26. Drechsler, E. R., Boyer, P. D. & Kowalsky, A. G. (1959). The catalytic activity of carboxypeptidase-degraded aldolase. *J. Biol. Chem.* **234**, 2627-2634.

27. Rutter, W. J., Richards, O. C. & Woodfin, B. M. (1961). Comparative studies of liver and muscle aldolase I. Effect of carboxypeptidase on catalytic activity. *J. Biol. Chem.* **236**, 3193-3197.

28. Susor, W. A., Kochman, M. & Rutter, W. J. (1969). Heterogeneity of presumably homogeneous protein preparations. *Science* **165**, 1260-1262.

29. Koida, M., Lai, C. Y. & Horecker, B. L. (1969). Subunit structure of rabbit muscle aldolase: extent of homology of the α and β subunits and age-dependent changes in their ratio. *Archs Biochem. Biophys.* **134**, 623-631.

30. Chan, W., Morse, D. E. & Horecker, B. L. (1967). Nonidentity of subunits of rabbit muscle aldolase. *Proc. natn. Acad. Sci. U.S.A.* **57**, 1013-1020.

31. Kowalsky, A. & Boyer, P. D. (1960). A carboxypeptidase-H_2O^{18} procedure for determination of COOH-terminal residues and its application to aldolase. *J. Biol. Chem.* **235**, 604-608.

32. Winstead, J. A. & Wold, F. (1964). Studies on the carboxyl- and aminoterminal residues of rabbit muscle aldolase. *J. Biol. Chem.* **239**, 4212-4216.

33. Morse, D. E., Chan, W. & Horecker, B. L. (1967). The subunit structure and carboxy-terminal sequence of rabbit muscle aldolase. *Proc. natn. Acad. Sci. U.S.A.* **58**, 628-634.

34. Penhoet, E., Rajkumar, T. & Rutter, W. J. (1966). Multiple forms of fructose diphosphate aldolase in mammalian tissues. *Proc. natn. Acad. Sci. U.S.A.* **56**, 1275-1282.

35. Lai, C. Y. & Chen, C. (1968). Studies on the structure of rabbit muscle aldolase II. Primary structure of peptide CnV and chemical evidence for the four-subunit structure of aldolase. *Archs Biochem. Biophys.* **128**, 212-218.

36. Lai, C. Y., Chen, C. & Horecker, B. L. (1970). Primary structure of two COOH-terminal hexapeptides from rabbit muscle aldolase: a difference in the structure of the α and β subunits. *Biochem. Biophys. Res. Commun.* **40**, 461-468.

37. Lai, C. Y. & Horecker, B. L. (1970). Modification of rabbit muscle aldolase *in vivo* and *in vitro*. *J. Cell. Physiol.* **76**, 381-388.

38. Midelfort, C. F. & Mehler, A. H. (1972). Deamidation *in vivo* of an asparagine residue of rabbit muscle aldolase. *Proc. natn. Acad. Sci., U.S.A.* in press.

39. Schapira, G., Kruh, J., Dreyfus, J. C. & Schapira, F. (1960). The molecular turnover of muscle aldolase. *J. Biol. Chem.* **235**, 1738-1741.

40. Udenfriend, S. & Velock, S. F. (1951). The isotope derivative method of protein amino end-group analysis. *J. Biol. Chem.* **190**, 733-740.

41. Edelstein, S. J. & Schachman, H. K. (1966). Studies on the polypeptide chains of rabbit muscle aldolase. *Fed. Proc. Fed. Am. Soc. Exp. Biol.* **25**, 412.

42. Lai, C. Y., Hoffee, P. & Horecker, B. L. (1965). Mechanism of action of aldolases. XII. Primary structure around the substrate binding site of rabbit muscle aldolase. *Archs Biochem. Biophys.* **112**, 567-579.

43. Lai, C. Y. & Oshima, T. (1971). Studies on the structure of rabbit muscle aldolase. III. Primary structure of the BrCN peptide containing the active site. *Archs Biochem. Biophys.* **144**, 363-374.

44. Sajgo, M. (1969). Distribution of the sulfhydryl groups of rabbit muscle aldolase in the polypeptide chain. *Acta Biochim. Biophys. Acad. Sci. Hung.* **4**, 385-389.

45. Perham, R. N. & Anderson, P. J. (1970). The reactivity of thiol groups in aldolase. In *Biochemical Society Symposia*, No. 31, Chemical reactivity and the biological role of functional groups in enzymes, pp. 49-58. (Smellie, R. M. S., ed.) Academic Press, London and New York.

46. Hartley, R. W. (1970). Derivatives of *Bacillus amyloliquefaciens* ribonuclease (barnase) isolated after limited digestion by carboxypeptidases A and B. *Biochem. Biophys. Res. Commun.* **40**, 263-270.

47. Morse, D. E. & Horecker, B. L. (1968). The mechanism of action of aldolases. In *Advances in Enzymology*, (Nord, F. F., ed.), vol. 31, pp. 125-181, Interscience Publishers, New York.

48. Steinman, H. M. & Richards, F. M. (1970). Participation of cysteinyl residues in the structure and function of muscle aldolase. Characterization of mixed disulfide derivatives. *Biochemistry* **9**, 4360-4372.

49. Eagles, P. A. M., Johnson, L. N. & Joynson, M. A. (1969). Subunit structure of aldolase: Chemical and crystallographic evidence. *J. Molec. Biol.* **45**, 533-544.

50. Szajani, B., Sajgo, M., Biszku, E., Friedrich, P. & Szabolcsi, G. (1970). Identification of a cysteinyl residue involved in the activity of rabbit muscle aldolase. *Eur. J. Biochem.* **15**, 171-178.

51. Anderson, P. J. & Perham, R. N. (1970). The reactivity of thiol groups and the subunit structure of aldolase. *Biochem. J.* **117**, 291-298.
52. Kobashi, K. & Horecker, B. L. (1967). Reversible inactivation of rabbit muscle aldolase by o-phenanthroline. *Archs Biochem. Biophys.* **121**, 178-186.
53. Lai, C. Y., Chen, C., Smith, J. D. & Horecker, B. L. (1971). The number, distribution and functional implication of sulfhydryl groups in rabbit muscle aldolase, *Biochem. Biophys. Res. Commun.* **45**, 1497-1505.
54. Suh, B. & Barker, R. (1971). Fluorescence studies of the binding of alkyl and aryl phosphates to rat muscle aldolase. *J. Biol. Chem.* **246**, 7041-7050.
55. Westheimer, F. H. & Cohen, H. (1938). The amine catalysis of the dealdolization of diacetone alcohol. *J. Amer. Chem. Soc.* **60**, 90-94.
56. Speck, Jr., J. C. & Forist, A. A. (1957). Kinetics of the amino acid-catalysed dealdolization of diacetone alcohol. *J. Amer. Chem. Soc.* **79**, 4659-4660.
57. Horecker, B. L., Pontremoli, S., Ricci, C. & Cheng, T. (1961). On the nature of the transaldolase-dihydroxyacetone complex. *Proc. natn. Acad. Sci. U.S.A.* **47**, 1949-1955.
58. Grazi, E., Cheng, T. & Horecker, B. L. (1962). The formation of a stable aldolase-dihydroxyacetone phosphate complex. *Biochem. Biophys. Res. Commun.* **7**, 250-253.
59. Model, P., Ponticorvo, L. & Rittenberg, D. (1968). Catalysis of an oxygen-exchange reaction of fructose 1,6-diphosphate and fructose 1-phosphate with water by rabbit muscle aldolase. *Biochemistry* **7**, 1339-1347.
60. Cash, D. J. & Wilson, I. B. (1966). The cyanide adduct of the aldolase dihydroxyacetone phosphate imine. *J. Biol. Chem.* **241**, 4290-4292.
61. Schmidt, Jr., D. E. & Westheimer, F. H. (1971). pK of the lysine amino group at the active site of acetoacetate decarboxylase. *Biochemistry* **10**, 1249-1253.
62. Castellino, F. J. & Barker, R. (1966). The binding-sites of rabbit muscle aldolase. *Biochem. Biophys. Res. Commun.* **23**, 182-187.
63. Ginsburg, A. & Mehler, A. H. (1966). Specific anion binding to fructose diphosphate aldolase from rabbit muscle. *Biochemistry* **5**, 2623-2634.
64. Ginsburg, A. (1966). The active sites of rabbit muscle aldolase. *Archs Biochem. Biophys.* **117**, 445-450.
65. Shapiro, S., Enser, M., Pugh, E. & Horecker, B. L. (1968). The effect of pyridoxal phosphate on rabbit muscle aldolase. *Archs Biochem. Biophys.* **128**, 554-562.
66. Yasnikov, A. A., Boiko, T. S., Volkova, N. V. & Mel'nichenko, I. V. (1966). On the mechanism of aldolase action. *Biokhimiya* **31**, 969-975.
67. Hoffee, P., Lai, C. Y., Pugh, E. L. & Horecker, B. L. (1967). The function of histidine residues in rabbit muscle aldolase. *Proc. natn. Acad. Sci. U.S.A.* **57**, 107-113.
68. Horecker, B. L. & Smyrniotis, P. Z. (1955). Purification and properties of yeast transaldolase. *J. Biol. Chem.* **212**, 811-836.
69. Westhead, E. W. (1965). Photooxidation with Rose bengal of a critical histidine residue in yeast enolase. *Biochemistry* **4**, 2139-2144.

70. Davis, L. C., Brox, L. W., Gracy, R. W., Ribereau-Gayon, G. & Horecker, B. L. (1970). Photosensitization of specific histidine residues of rabbit muscle and spinach leaf aldolases by pyridoxal phosphate. *Archs Biochem. Biophys.* **140**, 215-222.
71. Davis, L. C., Ribereau-Gayon, G. & Horecker, B. L. (1971). Photoinactivation of aldolases by pyridoxal phosphate and its analogues. *Proc. natn. Acad. Sci. U.S.A.* **68**, 416-419.
72. Rippa, M. & Pontremoli, S. (1969). Pyridoxal 5'-phosphate as a specific photosensitizer for histidine residue at the active site of 6-phosphogluconate dehydrogenase. *Archs Biochem. Biophys.* **133**, 112-118.
73. Mehler, A. H. & Cusic, Jr., M. E. (1967). Aldolase reaction with sugar diphosphates. *Science, N.Y.* **155**, 1101-1103.
74. Rose, I. A., O'Connell, E. L. & Mehler, A. H. (1965). Mechanism of the aldolase reaction. *J. Biol. Chem.* **240**, 1758-1765.
75. Spolter, P. D., Adelman, R. C. & Weinhouse, S. (1965). Distinctive properties of native and carboxypeptidase-treated aldolases of rabbit muscle and liver. *J. Biol. Chem.* **240**, 1327-1337.
76. Schmid, A., Christen, Ph. & Leuthardt, F. (1966). Uber die Reaktion der Muskel- und Leber-Aldolase mit Dansylchlorid und Acetylimidazol. *Helv. Chim. Acta* **49**, 281-287.
77. Pugh, E. L. & Horecker, B. L. (1967). The effect of acetylation on the activity of rabbit muscle fructose diphosphate aldolase. *Biochem. Biophys. Res. Commun.* **26**, 360-365.
78. Pugh, E. L. & Horecker, B. L. (1967). Function of tyrosine residues in rabbit muscle aldolase. *Archs Biochem. Biophys.* **122**, 196-203.
79. Swenson, A. D. & Boyer, P. D. (1957). Sulfhydryl groups in reaction to aldolase structure and catalytic activity. *J. Amer. Chem. Soc.* **79**, 2174-2179.
80. Kowal, J., Cremona, T. & Horecker, B. L. (1965). The mechanism of action of aldolases. IX. Nature of the groups reactive with chlorodinitrobenzene. *J. Biol. Chem.* **240**, 2485-2490.
81. Lai, C. Y. Martinez-de Dretz, G., Bacila, M., Marinello, E. and Horecker, B. L. (1968). Labeling of the active site of aldolase with glyceraldehyde 3-phosphate and erythrose 4-phosphate. *Biochem. Biophys. Res. Commun.* **30**, 665-672.
82. Wagner, J., Lai, C. Y. & Horecker, B. L. (1972). The nature of the aldehyde binding site of rabbit muscle aldolase. *Archs Biochem. Biophys.* **152**, 398-403.
83. Gray, G. R. (1971). An examination of D-fructose 1,6-diphosphate and related sugar phosphates by fourier transform ^{31}P nuclear magnetic resonance spectroscopy. *Biochemistry* **10**, 4705-4711.
84. Anai, M., Lai, C. Y. & Horecker, B. L. (1972). The site of pyridoxal phosphate binding in rabbit muscle aldolase. (Manuscript in preparation).
85. Leuthardt, F., Testa, E. & Wolf, H. P. (1952). Uber den Stoffwechsel des Fructose-1-phosphats in der Leber. *Helv. physiol. pharmacol. Acta* **10**, c57-c59.
86. Hers, H. G. & Kusaka, T. (1953). Le Metabolisme du Fructose-1-Phosphate dans le Foie. *Biochim. Biophys. Acta* **11**, 427-437.
87. Peanasky, R. J. & Lardy, H. A. (1958). Bovine liver aldolase. I. Isolation, crystallization, and some general properties. *J. Biol. Chem.* **233**, 365-370.

88. Penhoet, E., Kochman, M., Valentine, R. & Rutter, W. J. (1967). The subunit structure of mammalian fructose diphosphate aldolase. *Biochemistry* 6, 2940-2949.

89. Lebherz, H. G. & Rutter, W. J. (1969). Distribution of fructose diphosphate aldolase variants in biological systems. *Biochemistry* 8, 109-121.

90. Penhoet, E. E. & Rutter, W. J. (1971). Catalytic and immunochemical properties of homomeric and heteromeric combinations of aldolase subunits. *J. Biol. Chem.* 246, 318-323.

91. Meighen, E. A. & Schachman, H. K. (1970). Hybridization of native and chemically modified enzymes. I. Development of a general method and its application to the study of the subunit structure of aldolase. *Biochemistry* 9, 1163-1176.

92. Ting, S.-M., Lai, C. Y. & Horecker, B. L. (1971). Primary structure at the active sites of beef and rabbit liver aldolases. *Archs Biochem. Biophys.* 144, 476-484.

93. Felicioli, R. & Horecker, B. L. (1972). The active site of rabbit brain aldolase (aldolase C). (Manuscript in preparation).

94. Lai, C. Y. & Chen, C. (1971). Codfish muscle Aldolase: Purification, properties, and primary structure around the substrate-binding site. *Archs Biochem. Biophys.* 144, 467-475.

95. Ting, S.-M., Sia, C. L., Lai, C. Y. & Horecker, B. L. (1971). Frog muscle aldolase: Purification of the enzyme and structure of the active site. *Archs Biochem. Biophys.* 144, 485-590.

96. Gibbons, I., Anderson, P. J. & Perham, R. N. (1970). Amino acid sequence homology in the active site of rabbit and sturgeon muscle aldolases. *FEBS Lett.* 10, 49-53.

97. Guha, A., Lai, C. Y. & Horecker, B. L. (1971). Lobster muscle aldolase: Isolation, properties and primary structure at the substrate-binding site. *Archs Biochem. Biophys.* 147, 692-706.

98. Ribereau-Gayon, G., Ramasarma, T. & Horecker, B. L. (1971). Fructose diphosphate aldolase of spinach leaf: Primary structure around the substrate-binding site. *Archs Biochem. Biophys.* 147, 343-348.

99. Kobes, R. D., Simpson, R. T., Vallee, B. L. & Rutter, W. J. (1969). A functional role of metal ions in a class II aldolase. *Biochemistry* 8, 585-588.

100. Harris, C. E., Kobes, R. D., Teller, D. C. & Rutter, W. J. (1969). The molecular characteristics of yeast aldolase. *Biochemistry* 8, 2442-2454.

101. Mildvan, A. S., Kobes, R. D. & Rutter, W. J. (1971). Magnetic resonance studies of the role of the divalent cation in the mechanism of yeast aldolase. *Biochemistry* 10, 1191-1204.

102. Ingram, J. M. (1967). Fructose diphosphate aldolase from *Fusarium oxysporum f. lycopersici.* III. Studies on the mechanism of reaction. *Can. J. Biochem. Physiol.* 45, 1909-1917.

103. Harris, I. (1972). Personal communication.

Author Index

Numbers followed by asterisks indicate the page on which the reference is listed. Numbers in brackets are reference numbers and are included to assist in locating references in which the authors names are not mentioned in the text.

H

Haestis, W. H., 95(14), 105*
Halac, E., 132(90), 133(103), 135(103), 145*, 146*
Halac, E. Jr., 132(89), 145*
Hall, C. W., 135(120), 147*
Hall, E. A., 50(54, 55), 74*
Halperin, M. L., 23(143), 34*
Hammaker, L., 114(16), 115(18), 118(16), 136(127), 141*, 147*
Hancock, I. C., 60(81), 68(102, 103), 69(102), 75*, 77*
Hanninen, O., 135(121), 147*
Hardman, K. D., 95(12), 105*
Hargreaves, T., 125(59), 137(138), 143*, 148*
Harris, C. E., 171(100), 172(100), 178*
Harris, I., 172(103), 178*
Harris, J. E., 13(90), 31*
Harris, R. K., 84(3), 105*
Hartley, R. W., 155(46), 175*
Hartmann, G., 6(26), 27*
Haslam, R., 124(50), 143*
Hawkins, R. A., 17(118), 32*
Hay, J. B., 38(5), 68(104), 71*, 77*
Haymaker, L., 125(56), 143*
Heckels, J. E., 49(48), 74*
Heidelberger, M., 48(43, 44, 45), 73*, 74*
Heinz, F., 9(66), 30*
Heirwegh, K. P. M., 123(49), 124(51, 52), 126(68), 128(82), 130(49, 82, 83), 132(52, 92, 93), 133(98, 101), 143*, 144*, 145*, 146*
Hemingway, E., 136(132), 147*
Hems, D. A., 24(146), 34*
Hems, R., 11(74), 17(121, 122), 19(122), 30*, 33*
Henderson, J. J., 6(28), 28*
Heptinstall, S., 40(10), 46(37), 49(48, 49, 50), 68(106), 69(106), 70(106), 72*, 73*, 74*, 77*
Herman, R. H., 13(82, 83), 30*, 31*
Hermann, L. S., 138(139), 148*
Hers, H. G., 9(65), 30*, 167(86), 177*
Hess, G., 20(128), 33*
Hiatt, H. H., 152(24), 174*
Hibbard, E., 138(144), 148*

Hiepler, E., 30*
Higashi, Y., 56(77), 61(77, 84), 65(92), 75*, 76*
Ho, W., 7(43), 28*
Hoffee, P., 154(42), 156(42), 161(67), 162(67), 169(42), 170(42), 175*, 176*
Hoffmann, H. N., II, 128(79), 144*
Hoffmann-Ostenhof, O., 14(99), 32*
Hofmann, A. F., 128(76, 77), 144*
Hopkins, F. G., 2(3, 4), 26*, 149(1), 173*
Horad, J. L., 108(1), 140*
Horecker, B. L., 150(7), 151(16, 17, 18), 152(20, 24, 29, 30), 153(29, 33), 154(36, 37, 42), 155(7, 36), 156(36, 42, 47), 157(52, 53), 158(16, 17, 57, 58), 159(18), 161(65, 67), 162(67, 68), 163(70, 71, 77), 164(52, 78, 80, 81, 82), 167(65, 84), 169(7, 42, 92, 93, 95, 98), 170(42, 92, 93, 95, 97), 173*, 174*, 175*, 176*, 177*, 178*
Houghton, C. R. S. 22(140), 33*
Housset, E., 136(126), 147*
Howe, R. B., 124(55), 143*
Hrkal, Z., 114(11), 141*
Hübscher, G., 13(81), 30*
Hughes, A. H., 68(102), 69(102), 77*
Hughes, N. A., 44(28), 73*
Hughes, R. C., 40(9), 45(34), 72*, 73*
Hulme, E. C., 7(42), 28*
Hussey, H., 59(80), 61(83), 63(83, 88), 64(91), 65(83), 66(83), 75*, 76*
Hutchinson, D. W., 121(41, 43), 122(41), 123(47), 124(41), 142*, 143*

I

Ibsen, K. H., 7(44), 28*
Illing, H. P. A., 130(84), 145*
Inagami, T., 95(12), 105*
Inciardi, N. F., 22(139), 33*
Ingram, D. J. E., 95(19), 106*
Ingram, J. M., 172(102), 178*
Ionesco, H., 67(93), 76*
Ishimoto, N., 54(70), 55(72), 68(70), 75*

Subject Index

A

Acetyl CoA, 8, 23, 24

α-N-Acetylglucosamine 1-phosphate, polymer of, 49

β-N-Acetylglucosaminidase, 54

N-Acetylglucosaminyl-N-acetyl-muramyl peptide pyrophosphate, 57, 62

N-Acetylmuramyl peptide monophosphate, 58

Acid haematin, 110

Acid phosphatase, 114

Actinomycetes, teichoic acids of, 43, 48

Active site,
  e.p.r. and n.m.r. studies of, 94-95, 98-100
  of aldolase, 158, 159-169

Adenine nucleotides,
  phosphorylation state of, 8
    Crabtree effect and, 21-22

Aerobic glycolysis,
  Meyerhof quotient and, 10-12
  nitrophenols and, 5
  physiological role of, 12-21
    in erythrocytes, 13-15
    in foetal tissues, 15
    in haemopoietic bone marrow cells, 16-17
    in intestinal mucosa, 12-13
    in malignant tumours, 15
    in renal medulla, 13
    in retina, 16-20
    in striated muscle, 12
  stimulation by ethylcarbylamine, 3

D-Alanine activating enzyme, 56

Alanine ester residues in teichoic acids, 56
  cation binding and, 68-69
  resistance to lysis and, 70

D-Alanine:membrane ligase, 56

Alanyl-AMP-enzyme complex, 56

Aldolase 9, 10, 149-173
  Class I,
    in plants, 169-171
    properties of, 150, 159
  Class II,
    properties of, 150, 159, 171-172
  distribution of, 150, 167-171
  enzyme-substrate complex, 150-151, 158-159, 164-166
  frog muscle, 169
  isoenzymes of, 167-169
  rabbit muscle,
    active centre,
      functional groups at, 159-164, 166, 167
        cysteines, 164
        histidines, 161-162
        lysines, 159-161, 167
        tyrosines, 163-164, 166
      hydrophobicity of, 158
      structure of, 164-166, 169
    amino acid sequences in, 152-153, 154-156, 169, 170
    3-dimensional structure of, 157-158
    isolation of, 151-152
    mechanism of action, 158-159, 164-167
    microheterogeneity of, 152-153
    molecular weight of, 152
    Schiff base intermediate, 151, 158-162, 164-166, 167
    substrate cleavage rates, 152
    subunits of, 152-154, 167, 168
      origin of, 153-154
    tertiary structure of, 154-158
    thiol groups in,
      classes of, 156-157
      positions of, 156-158
  spinach, 163, 169-171
  yeast, 171-172

Alkaline haematin, 110

Allosteric model, 97-98